FROM INTENT TO IMPLEMENTATION

BY
MITCH SCHNEIDER

THOMSON
DELMAR LEARNING

Australia • Canada • Mexico • Singapore • Spain • United Kingdom • United States

THOMSON
DELMAR LEARNING

Automotive Service Management

From Intent to Implementation

Mitch Schneider

Executive Director:
Alar Elken

Executive Editor:
Sandy Clark

Automotive Product Development Manager:
Timothy Waters

Development:
Kristen Shenfield

Executive Marketing Manager:
Maura Theriault

Channel Manager:
Elizabeth Lutz

Marketing Coordinator:
Brian McGrath

Executive Production Manager:
Mary Ellen Black

Production Manager:
Larry Main

Production Editor:
Elizabeth Hough

Editorial Assistant:
Kristen Shenfield

Cover Design:
Julie Lynn Moscheo

Printed in Canada
1 2 3 4 5 XX 06 05 04 03 02

For more information contact Delmar Learning
Executive Woods
5 Maxwell Drive, PO Box 8007,
Clifton Park, NY 12065-8007
Or find us on the World Wide Web at
http://www.delmar.com

Library of Congress Cataloging-in-Publication Data:

Schneider, Mitch.
Automotive service management. From intent to implementation / Mitch Schneider.
p. cm.
ISBN 1-40182-658-X
1. Service stations--Management. 2. Business planning. I. Title: From intent to implementation. II. Title.
TL153 .S333 2003
629.28'6'068--dc21

2002151916

NOTICE TO THE READER

Publisher does not warrant or guarantee any of the products described herein or perform any independent analysis in connection with any of the product information contained herein. Publisher does not assume, and expressly disclaims, any obligation to obtain and include information other than that provided to it by the manufacturer.

The reader is expressly warned to consider and adopt all safety precautions that might be indicated by the activities herein and to avoid all potential hazards. By following the instructions contained herein, the reader willingly assumes all risks in connection with such instructions.

The publisher makes no representation or warranties of any kind, including but not limited to, the warranties of fitness for particular purpose or merchantability, nor are any such representations implied with respect to the material set forth herein, and the publisher takes no responsibility with respect to such material. The publisher shall not be liable for any special, consequential, or exemplary damages resulting, in whole or part, from the readers' use of, or reliance upon, this material.

CONTENTS

PREFACE

This shop operations guide is designed to provide a framework that helps you make consistent, high-quality, and productive service a part of your normal, everyday shop operations.

Is it everything you ever wanted to know about automotive shop management but were afraid to ask?

I'm not sure. In reality, it is a lot of what I have learned about running a successful automotive service business, both on my own and sitting at the feet of those who started this journey before I did. At the very least, it's everything you can fit into eight volumes.

The purpose of this guide is to help ensure that *great performance* is achieved every time you or one of your employees approaches a regular or potential customer to deliver automotive service. It is the only way I know to create trust and insure customer loyalty. And, great performance coupled with increased customer loyalty, trust, and operational excellence will almost always result in increased profits.

High-quality service is defined by the customer's experience while in your care and custody. Operational excellence ensures that the customer's cars will be *ready when promised* and *fixed right the first time.* Fixing the car right the first time is as much about understanding what your customer is trying to tell you when they are communicating their frustrations with the vehicle as it is about proper diagnosis and professional quality work. *Total Quality Service* is only possible through high-quality communication and knowing where you are in the process of completing a job at all times. Because the workforce is diverse, we chose to alternate between "he" and "she" from chapter to chapter to ensure gender is equally represented.

Selling service is perhaps the most difficult of all sales to make. Buying a product—something you can hold in your hand like an oil filter or a spark plug—constitutes the purchase of something you can touch or feel. Purchasing a service is something else entirely. When a vehicle owner buys a service, that person is really purchasing a promise—a promise that will be fulfilled in the future. That requires trust and a great deal of faith.

Your ability to perform is based upon a lot of things, not the least of which is technical competency. But you can't demonstrate that technical competency unless (or until) you are given the opportunity. The vehicle owner can't see technical competency. She can't touch it or feel it and you can't hang it out in front of the building like a sign.

The customer's perception of whether or not you are worthy of her trust, and whether or not you will be able to solve her automotive service problem, will be based in many cases upon any one of a number of subtle, almost subliminal, messages. These messages are broadcast and received before any one of your technicians ever picks up a screwdriver, wrench, ratchet, or socket.

Service starts when a customer sees and responds to an advertisement (an invitation to experience the quality of your service), hears about your business from a friend or relative, calls, or walks in, service includes every subsequent customer contact until the job is completed—ready when promised. Moments of truth—moments when a customer has the opportunity to come in contact with any member of your organization and form an opinion about you or your company—occur before, during, and after the job has been completed. They will ultimately define your personal success and the success of your company.

Consistent execution of company policies and procedures is the catalyst for world-class service. It is perhaps more important to be consistent at what you do than to be good at what you do. Do a great job one time, and something less than great the next, and you aren't likely to see that customer again. The more people you have working for you, the greater the challenge is to achieve consistency. Without a comprehensive and complete set of policies and procedures, consistency is nothing more than an elusive dream.

All of this consistent high-quality service is only sustainable if your shop is operating at productivity levels that allow you to make a profit. Productivity is a by-product of your technicians' ability to execute policies and procedures flawlessly. It will take more than a visit to the tool truck or a one-day management seminar your technicians to bill 70 percent or more of their available hours.

All eight volumes of this guide—*Total Customer Relationship Management, From Intent to Implementation, Managing Dollars With Sense, Operational Excellence, Building a Team, The High Performance Shop, Safety Communications, and Operations Management*—are designed to work together in large shops and small shops. That means that the building blocks of consistency—the policies and procedures you will find within these pages—are defined in their simplest terms because no matter how big or how small a shop is, the same kinds of tasks must be performed. The policies and procedures we define here can and will be carried out by the service writer, service manager, technician, office manager, and lot person. In many cases, the person who fills the role of service writer also serves as the service manager and perhaps even the technician. The key to success is not who does the job. The key to success is insuring that each critical task is completed accurately and consistently—the same way every time.

While this set of guides will take you through a series of step-by-step procedures, it is important to remember that delivering high-quality service is a process. To achieve success, all of the steps in the process must be performed, either consciously or unconsciously. If you are having a problem in one particular area of your business, start with the guide that addresses that need first. Next, read through the rest of the guides. See if what you find in these pages makes sense to you. *Then, read it again*. Look at your business and see what, if anything, applies to you. *Then, read it again*. Consider which of your current procedures need to remain in place. If some need to be changed, determine if some of the policies or procedures you found in these guides can be incorporated into the daily operation of your business. *Then, read it again*. Put yourself in the place of your customer. Involve everyone in your business in this exercise. You will be far more successful with their help and cooperation than you could ever be without it. Help them understand that this is how things need to be done from *this moment forward* and why. Walk through the new procedures in your shop. Pull the trigger, start the process, read the text, and begin your journey to a better quality of service, a stronger, healthier, more profitable business, and a more satisfying, less frustrating life.

ACKNOWLEDGEMENTS

Nothing of consequence occurs in a vacuum. It takes a lifetime of collaboration, cooperation, and support. The Automotive Service Management series is no exception. If it takes a village to raise a child, it takes a city to conceive, write, edit, and publish one book—let alone eight. Consequently, there are some special people I would like to thank for making the journey with me.

I'd like to thank my Delmar family: Tim Waters, Kristen Shenfield, Betsy Hough, and everyone in marketing for staying with me from conception to completion—as well as Lynanne Fowle and Megan Iverson at TIPS Technical Publishing for their meticulous attention to every detail. It is a joy to work with so many people who care.

I'd like to thank John Wick and Willi Alexander for planting the seeds, and Dick Vinet, the Duffy-Vinet Institute, Motor Service Magazine, and Champion Spark Plugs, for presenting the first shop management seminar I ever attended. It gave me confidence, but more than that—it gave me hope. And, Arthur Epstein (of blessed memory) and Ray Walker for their guidance, friendship, and patience and for recognizing in me something I could not or would not see myself.

Most of all I'd like to thank my wife, Lesley, for tolerating a lifetime of ruined dinners, working too late at the shop, and hiding out in the "Cave" writing and thinking and just generally being far away. She's not only been the biggest part of everything I've ever accomplished, believing even when I didn't—she's been the best part of everything I've ever accomplished. I'd like to thank my kids, Ryan and Dana, for sharing me with my work even when I'm sure they didn't understand. They are a blessing and an inspiration.

I'd like to thank my brothers: David for working with me for the better part of thirty-six years, and Todd for not. Both have taught me a great deal. David has taught me about patience and understanding and looking beyond my frustration to try to understand the lesson he is trying to teach me about myself. And, Todd has taught me a great deal about positive attitude, professionalism and personal courage, in addition to never saying No to anything I've ever asked of him.

Finally, I would like to acknowledge my parents—Sylvia and Jerry Schneider. They are far more than parents. They are my closest friends, my favorite teachers, and a living example of what excellence, character, commitment, dedication, sacrifice, and giving everything you have to give is all about. It's hard to relax and take it easy when your eighty-year-old parents won't, and it's just as hard to give less than your best when your eighty-year-old parents won't!

INTRODUCTION

From Intent to Implementation: Vision to Reality... Moving Closer to Your Dreams

"The reasonable man adapts himself to the world; the unreasonable man persists in trying to adapt the world to himself. Therefore, all progress depends on the unreasonable man."

George Bernard Shaw

The great bulk of humanity seems content with whatever the next day brings. They wake up in the morning, have their tea or coffee, read the paper, get dressed and then shuffle off to work. They punch a time clock or function in mid-management positions, build cabinets or install windows, program computers or stock boxes in a warehouse somewhere. In most cases, they work for people who have invested their life-savings or the money entrusted to them by their stockholders in a business or corporation, and that's fine. They are the majority, the great unwashed, the average guy or gal on the street. And, for most of them, one day looks and feels remarkably like the one before or the one after.

Everyone has dreams. Average people dream about a better job, a bigger house, a fatter paycheck, more time off, greater responsibilities at work or sometimes, even fewer responsibilities at work. They dream about a different reality, a better reality—and that's where it stops... with their dreams.

There are those who do more than dream, however—individuals cut from a different cloth... people who dream about a different tomorrow... a better tomorrow. Then, they do something about those dreams. They act!

They take their dreams and use them as a target—the ultimate target. They determine exactly what it might take to turn those dreams into reality, the resources it might require, the time and energy involved, and then they break those down into concrete goals and objectives. They give those goals and objectives a deadline and then they dig down, take a deep breath and move one step closer to their dreams every time they take a step.

What they do is often unwise, rarely prudent, difficult to explain or articulate and impossible to understand, especially for the majority of those among us unable or unwilling to make the journey with them.

This book is all about what it takes to move from dreams to reality. It is all about the steps you will need to take and the technologies that will be required. In the end, it is meant as a kind of roadmap for those who would like to do more than dream about tomorrow—those who would rather create their own epic production of tomorrow rather than remain an extra in someone else's.

In the end, it is about moving from the intent to do something different—something great—to the implementation of the actual systems, policies, and procedures required to do it.

T.E. Lawrence said that

> *"All men dream but not equally. Those who dream by night in the dusty recesses of their minds awake to find that it was vanity; But the dreamers of day are dangerous men, that they may act their dreams with open eyes to make it possible."*

Ultimately, the choice is yours. Will your dreams remain nothing more or less than unfulfilled dreams? Or, will you act upon your dreams with open eyes to make them possible?

CHAPTER 1

Why Are You Reading This Book?

INTRODUCTION

This chapter will explore:

- Your reasons for reading this book
- The concept of paradigms
- The reality that two people can look at the same thing seeing something entirely different based upon their experience and understanding
- Your paradigms regarding this industry
- Our resistance to change
- The relationship between vision and current reality and the impact that can have on change

SURVIVAL OR SUCCESS

Problems cannot be solved with the same level of awareness that created them...

Albert Einstein

If you are anything like I am, your time is at the very least *limited*. With the pressures of school, work, family, relationships, and recreation abundant in each of our lives, time has become one of the most valuable commodities—if not *the* most valuable commodity in our lives. There are certainly more demands on our time than there is time to meet those demands. So, I have to ask you: Why are you here? Why are you reading this book?

Although I can't think of a better, more productive way to spend your time, there are certainly a number of other things you could be doing that are at the very least, far more entertaining. And, yet, here you are spending the morning, afternoon or evening with me as we both embark upon a shop management odyssey. It will take us all the way from the realization that where you are is not where you want to be, all the way to wherever *where you want to be* might take you.

We are going all the way from intent (the decision that we must do something) to implementation (moving through the individual steps required to get it done)!

So why don't you take a moment to identify those factors motivating you to spend your time with me rather than in the pursuit of some other goal or objective?

Are you interested in having:

1. More time to spend with your family?
2. A better quality of life?
3. More money?
4. Greater security?
5. Less stress?
6. More time away from the shop?
7. Less responsibility?

Why did you make the commitment? What exactly is it you are looking for? What kind of a return do you hope to enjoy from the investment in time, effort and energy you are about to make?

Before you answer, consider this. Most automotive repair facilities across the nation show net profit results of five percent or less. Most shop owners and managers do not have a clue how well or how poorly they are doing, how hard they're working, or even if they are making any money at all!

If you ask the owners or managers of those facilities if they have a *plan*, they will probably look at you, smile wryly and mutter "Sure, survival!"—when the fact of the matter is, they aren't surviving no matter how close they think they're coming to it!

There is a direct correlation between the shop owners and managers who think about more than just getting through the day and success. There is a direct correlation between the shop owners and managers who are able to conceive of a new and different, powerful and positive vision for the future and those who cannot—or just plain won't. The shops guided by a vision and the mission to make that vision a reality, combined with the goals and objectives that spin out of that mission, are the ones showing above average growth and above average profits. The only question then, is: Do you want to be among them? If so, buckle your seat belt and hang on because we are headed for the fast lane on the strategic planning superhighway...

Why are you here?

1.
2.
3.
4.
5.

Perception is Reality... Feelings are Facts!

Your perceptions of reality *are* reality for you. However, is your reality the world as it actually is or is it the world as you see it filtered through your assumptions, experiences, and expectations?

per·cep·tion (pər-sep'shən) *noun* 1. **a.** Insight, intuition, or knowledge gained by perceiving. **b.** The capacity for such insight.

These experiences and expectations create a model of the world that helps us understand and make sense of what we encounter. Those models are referred to in most business management texts as paradigms. According to Joel Arthur Barker in his book *Future Edge,* "A paradigm is a set of rules and regulations (written or unwritten) that does two things: 1) it establishes or defines boundaries; and 2) it tells you how to behave inside those boundaries in order to be successful."

Paradigms are the world you experience colored by your assumptions about that world. Adam Smith suggests in *Powers of the Mind* that these assumptions are required to make sense of the world around us: "The paradigm explains the world to us and helps us to predict behavior."

In other words, it is our paradigms that help us tie all the loose ends together—connect the dots even when there are dots missing—allowing us to make sense of an otherwise random and chaotic world. In the end, they determine what we see and what we don't see, just as much as they influence our understanding and govern our reactions.

Thomas Kuhn, physicist and father of the concept of the paradigm, suggested that "You don't see something until you have the right metaphor to let you perceive it." In other words, the world is the world as we think we see it, not as it actually is. So if we change our paradigms—the lenses through which we see that world—we change the world itself, or at least how we will operate in that world.

The purpose of any exchange of ideas such as this one is to expand your paradigms and to modify, if not change, your assumptions. As Kuhn suggested in his book, *The Structure of Scientific Revolutions*, this transition might not reflect "a steady, cumulative acquisition of knowledge." It might, in fact, be "punctuated by intellectually violent revolutions" in which "one conceptual world-view was (is) replaced by another." It is these world-views that we have come to know as *paradigms*.

If I am successful, you will see yourself, your business, and those around you in a very different way, and because of that change in perception, you will act and react differently from that moment on.

What Do You See?

Since each of us is unique in so many ways, with very different expectations and experiences, our paradigms can be very different as well. Because our paradigms are different, so are our perceptions. We change our lives by changing our attitudes, assumptions, and perceptions—by changing our paradigms. We try to achieve behavior that is consistent with our self-image. In essence, we try to become what we think about ourselves. However, there is a problem. Since perception is colored by experience and assumption, no two people see the world precisely the same way.

Two people can look at the same event, even the same illustration, and see two completely different things. What do you see when you look at the picture in Figure 1-1?

Figure 1-1 Ugly woman (Source: Stephen Covey, 7 Habits of Highly Effective People)

Do you really see an ugly old woman? Or, do you see an elegant young lady? (See Figure 1-2.)

Figure 1-2 Elegant young lady.

You See What You See...

You see what you see because of *who* you are, *where* you have been, *what* you have done, and because of *what* you want to see and what you know. But, once again, is what you know based upon the world as it is? Or, is it based upon the world as you would like to see it... as you would like it to be?

> *Every business, in fact, every organization, operates on... a set of assumptions (paradigms) regarding the outside world (customer, markets, distributive channels, competition) and a set of assumptions regarding the inside (core competencies, technology, products, process). These assumptions are usually taken for holy writ by the company and its executives. It is on them that they base their decisions, their actions, their behavior.*
>
> *Peter Drucker*

What are your paradigms—your *assumptions*—about this industry? What are your paradigms with regard to your customers? What are your paradigms about your business or your relationship with your technicians or other employees?

Are you aware of just what those paradigms are?

Just because you may not be aware of them doesn't mean they don't exist! Each of us brings a set of assumptions about virtually everything we have experienced in the past, or may experience in the future, with us as we travel through the world. As we have already said, these assumptions are based upon our understanding of the world as we have encountered it and colored by the expectations we acquire as a result—in many ways, they define the boundaries of what we perceive to be possible.

Your paradigms about business—your business in particular (employees, income, wealth, customers, and customer service)—will define your business in virtually every way possible. It would be a good idea to know what they are. They will either chain you to the ground or give your success wings.

What are your paradigms about this industry?

1.
2.
3.
4.
5.

Paradigm Shifts

Mental models (paradigms) shape the way we organize information. They don't determine what information we have, they determine the sense we make out of that information.

Peter Senge

Change your paradigms and you change everything because a quantum shift in your belief system allows you to see new and different possibilities—new and different opportunities. It's not that they weren't there before; it's just that you never saw them. Or, perhaps, you saw them, but never had the technology to take advantage of them.

The problem with our paradigms is that they are comfortable, or at least allow us to feel comfortable even when we find ourselves in difficult or unfamiliar circumstances. They give us a certain level of confidence, whether real or imagined, that allows us to function. Our paradigms allow us to solve problems and interpret the world around us as long as that world stays pretty much the same. But, what happens when the world around us begins to change faster than our paradigms can keep up? What happens when the frame of reference we have developed to help us solve problems no longer works on the problems we must confront and then solve?

I'm not exactly sure if the world around us is changing quickly because that change is being driven by technology, or if technology is changing quickly in response to a rapidly changing world. Either way, everything that we have used to achieve whatever success we have accomplished to this point will not be enough to guarantee our success in the future. The world is changing. Technology is changing. The problems are changing. And consequently, our solutions are going to have to change as well.

There is only one problem—no one likes change. Almost everyone you know or have ever known hates it! In fact, someone once told me that there were only two organisms on the planet that welcome change—a stockbroker and a wet infant!

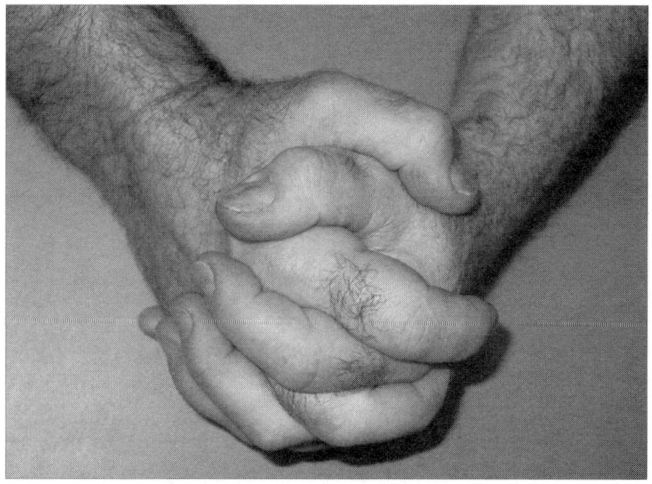

Figure 1-3 Change can be uncomfortable.

It's easy to understand why no one is willing to embrace change—it's uncomfortable! If you don't believe me, try this little experiment. Do you remember kindergarten? Remember when you learned to sit with your hands folded on top of your desk with your fingers interlaced? I'd like you to take a moment and try that for me right now. Interlock your fingers and rest the ridges of both hands on the top of your desk or table now. Take a minute to notice that one thumb has found its place on top of the other thumb, as seen in Figure 1-3.

Quite frankly, it really doesn't matter which thumb it is—you're going to find one on top.

Figure 1-4 Change can be difficult and uncomfortable.

Now, unlock your fingers and shake your hands out. I'd like you to interlock your fingers again, but this time place your *other* thumb on top. Give it a try... If you are like the thousands of people I've tried this with in the past, you just experienced how uncomfortable change can be. You have been interlocking your fingers like that since you were a child. I just asked you to break a habit—a pattern that is deeply established in both your mind and in your muscle memory. That's not an easy thing to do. It's not easy and it's not comfortable.

If you aren't quite sure about the previous experiment or its outcome, try this—extend both your arms out, fingers pointed forward, and then cross your arms one over the other in front of your body, as seen in Figure 1-4. Take a look at both your hands. One of them has found its way *under* the opposite bicep, the other has found its way *over*.

Take your arms and unfold them. Then, fold them again. Only this time, place the hand that was on top of the bicep under it and the one that was over, under it. I still can't do it!

Recognizing that change is difficult and uncomfortable is important. It helps us to be a little more understanding when those around us are reluctant to embrace the change that we know is necessary to create a new and different future for us all. It is equally important to recognize that we shouldn't be too quick to condemn those around us because they won't change. We often aren't quite ready or willing to embrace those changes ourselves!

Change & Movement

Moving from intent to implementation is only a matter of understanding where you are, recognizing where you want to be, establishing the distance between those two points, determining what it will take to bridge that gap, and then doing whatever is necessary to move from where you are to where you want to be. If it's that easy, then why do so many people seem to have so many problems moving from where they are to where they need to be?

There are lots of reasons—perhaps as many as there are people needing to make the journey. Some people are afraid of failure. Others are more terrified of success. Some people don't know how or where to begin, and others aren't exactly sure of what they need to do to get started. The excuses are endless and, like bellybuttons, everybody has at least one! But that doesn't mitigate the need for change.

Someone once said that there are three kinds of people in the world. One group is comprised of the people that make things happen. Another is comprised of the people that are comfortable just watching things happen. The final group is made up of all those individuals who wait and watch and wonder if something *just* happened. The world is changing and the choices are clear—you can lead the change, accept and participate in the changes led by others, or find yourself lost and falling behind in a world that has changed without your even being aware of it.

Change is all about ***character***—the ability to carry out a worthy decision after the emotion of making that decision has long since passed.

> **char·ac·ter** (kar'ək-tər) *noun* 1. The combination of qualities or features that distinguishes one person, group, or thing from another. See synonyms at disposition. 2. Moral or ethical strength. 3. A description of a person's attributes, traits, or abilities. 4. Public estimation of someone; reputation.

Character is one of those value words—like integrity, honor, trust, loyalty, generosity, compassion, and ethics—that describe traits we should all aspire to. Values are things we believe in and the things we believe in to a large degree are the foundations of some of our most important paradigms. Understanding values and recognizing how important they can be to both our success and our state of mind, and realizing the danger of taking actions that are inconsistent with those values, will play a major role in this odyssey.

Vision & Current Reality

A moment ago, we said that moving from intent to implementation is only a matter of understanding where you are, recognizing where you want to be, establishing the distance between those two points, determining what it will take to bridge that gap, and then doing whatever is necessary to move from where you are to where you want to be. Peter Senge talked about this kind of movement in terms of vision and current reality in *The Fifth Discipline*.

My sense of what vision is all about is fairly simple. It is what you see when you close your eyes and think about the way things could be—*should* be—in your conception of a perfect world. It is not the way things are—it's the way they ought to be... the way you would like them to be.

Vision is personal. It starts with you. It begins with *your* wants, needs, and expectations for the future, your future, and the future of those close to you. Vision is not restricted to a *corporate* boardroom or a commercial setting. In fact, in my opinion it would be difficult, if not almost impossible, to find yourself committed to establishing a corporate vision (a vision for your business) if you were unable or unwilling to create your own personal vision first.

After you have developed or established a personal vision, it can expand outward to include your personal circle of immediate relationships and then out to your business and your business relationships.

Your dreams, aspirations and hopes are the stuff vision is made of—and dreams, aspirations, and hopes are not restricted to the exclusive realm of business. In fact, they should originate in your personal life and then spill over into the world of business, raising the quality of your current reality—your life as it is, not your life as it could be—in both.

Figure 1-5 Creative tension.

For most of us, vision and current reality are in conflict, as seen in Figure 1-5. Peter Senge refers to this gap between vision and current reality as *creative tension.* As seen in Figure 1-6, creative tension is the tension (pressure, desire, anxiety, and anticipation) you feel when you know that where you are is not where you want to be, but realize that getting to where you want to be will require effort, exertion, determination, commitment, discipline, or even pain.

There are basically three dynamics that can and will occur once you recognize or establish your vision and realize the limits of your current reality. If your vision is too high—if it is virtually unattainable—or your current reality is too strong because of poor self-image, low self-esteem, a lack of confidence, or a sense of powerlessness or unworthiness, then the creative tension resulting from this gap will force you to lower your standards or it will bring your vision more in line with your current reality. If your vision is strong and vivid enough to overcome your sense of powerlessness or unworthiness, it can and often will, drag your current reality forward as you move closer to realizing that vision. And finally, if both your current reality and your vision are equally powerful, it is conceivable that the creative tension that occurs as a result may just cause you to snap!

Figure 1-6 Achieving your vision takes effort, exertion, determination, commitment, and discipline.

Reconciling the effort and pain of setting goals and attempting to achieve your vision with the possible risk of failure and the satisfaction of reward is what progress is all about. But the forces holding us back can be equally as powerful as the forces driving us forward. In addition to problems with poor self-image and low self-esteem, which are endemic in our industry, we continually fight a crisis of confidence—both internally and externally.

Vision drives the establishment of goals. Goals drive the establishment of objectives (tasks). Objectives demand action.

SUMMARY

All of us are products of a combination of experience and our own unique genetic cocktail. Our perceptions and assumptions (paradigms) are the basis of our beliefs and color the way we see the world. They are the lenses through which we see that world. Change the lenses and you can change the world.

What are your assumptions about this industry? What are your assumptions about your business and your ability to succeed? How would you define that success? What is your vision for the future and what is standing in the way of you achieving that vision?

Take a minute to write down some of the ways you think your assumptions—your paradigms—help you to deal with your current reality and some of the ways they get in the way. Use the space on the next page.

Thoughts

Actions

Results

CHAPTER 2

Taking Inventory

INTRODUCTION

This chapter will help you to recognize who you are in relation to your:

- Roles
- Goals
- Objectives
- Tactics
- Intentions

as well as where you are in regard to your professional life by looking at your:

- Work
- Income
- Freedom
- Family
- Time

and the difference between:

- Principles and values

Chapter 2

A SNAPSHOT OF WHO YOU ARE

If vision begins with the individual and then expands to the organization, it becomes critical that we understand the role your personal principles, values, roles, and objectives play in driving us toward success and keeping us on course. In order to do that we have to identify what they are and how they impact and color everything we do.

Principles

> *Principles are guidelines for human conduct that are proven to have enduring, permanent value. They're fundamental. They're essentially unarguable because they are self-evident.*
>
> *Stephen Covey*

prin·ci·ple (prin'sə-pəl) *noun Abbr.* prin. **1.** A basic truth, law, or assumption: *the principles of democracy.*

I believe that man is the only organism on the planet capable of self-awareness. It is a belief I share with Stephen Covey, the author of *7 Habits of Highly Effective People.* I don't mean *spatial self-awareness*—the kind of awareness that tells us where we are in the physical world. I mean a kind of *spiritual awareness* that allows us to consider what we are doing here and what our responsibilities are to ourselves and, perhaps more importantly, to each other. This self-awareness leads us to a set of *principles* that govern human interaction and effectiveness. These principles are irrefutable and unchanging, no different than the natural laws that govern our physical world, like gravity. The law of gravity exists whether you believe in it or not.

These principles are the foundation of virtually every civilized culture that has ever existed. Without them, man reverts back to something almost prehistoric, something somehow less civilized and somehow less human.

These principles are sewn into the genetic thread of our species. With our ability to consider more than just our physical needs, they separate us from the rest of the creatures that share the planet with us. I would argue that there are species that share just about every other characteristic and trait we selfishly call our own, such as the ability to show warmth to or experience a deep emotional bond with another, the instinct to protect our mate or our offspring, or the need to belong or experience companionship. But humans are the only species that I am aware of that can consider those feelings—those actions and reactions—contemplate them, articulate them, and then share them.

I don't think dogs think about their place in the universe or wonder about what happiness is. Man does. As a result, we live our lives based on certain principles, without which civilization could not exist—principles like *fairness, integrity, honesty, service to others, excellence,* our own *human potential,* and more.

Values

val·ue (val'yoo) *noun Abbr.* val. 1. A principle, standard, or quality considered worthwhile or desirable

Values are something you believe in. If those values are based on the kind of principles we spoke of above, they are useful and healthy. They allow us to negotiate our way through the world by helping us recognize what is significant and worthy in our lives.

A value is not the same thing as an objective. An objective is something you hope to attain or achieve. If *honesty* is the principle we hold fast to, then *truth* is the value we will hold dear. Unfortunately, it is possible to have values that do not flow out of principle. Crime families and criminals can share values; terrorists and fanatics can share values as well. They just aren't the values most of us would aspire to call our own.

Roles

role (rōl) *noun* 1. The characteristic and expected social behavior of an individual.

Each of us finds ourselves wearing many hats in the course of the day, month, year, or even lifetime. We are sons or daughters first, fathers and mothers later on. We are brothers or sisters, aunts or uncles, husbands or wives, employees or employers, and with each new and different *role* there comes a new set of expectations and responsibilities.

It is important to understand what these roles are, just as it is important to identify the responsibilities and expectations that accompany them. We will judge ourselves and be judged by others on how well we meet those responsibilities and exceed those expectations.

Goals

goal (gōl) *noun* 1. The purpose toward which an endeavor is directed; an objective.

As I mentioned earlier, a *goal* is something you hope to attain or achieve—something you work *for* or *toward.* Goals are the target. They determine what your *strategic objectives* will be. Strategic objectives are achieved through smaller bites we'll call tactics.

If what you are doing is not moving you closer to those goals or objectives (with regard to any one of the many roles you find yourself playing)—if it is in conflict with your values or your principles—you have to ask yourself why you are taking the time and making the effort to do it.

Objectives

ob·jec·tive (əb-jək'tiv) *noun* 1. Something worked toward or striven for.

Objectives are the steps you take to move closer to your goals. They are greater than the tactics that allow you to attain those objectives. If one of your goals is financial security, one of your strategic objectives might be a significant 401k or IRA that matured simultaneously with your retirement. A sound retirement strategy would require you to tactically deposit something in that account weekly, monthly, or at least yearly in order to reach that goal. Unless you plan on winning the lottery, achieving financial security is not something that will occur overnight! It requires all the *D* words everyone seems to have so much trouble with— drive, determination, and discipline.

It is important to realize that you can have a number of goals and even more objectives. You can strive for financial security, hope to own your own home, work for a new car or truck, or you can dream of owning a business of your own. The key is understanding the role principles and values play, or should play, in helping us choose the right goals and objectives.

If we fail to do that, we could find ourselves leading the pack, but running in the wrong race!

Tactics

tac·tic (tak'tik) *noun* An expedient for achieving a goal; a maneuver.

There are always a couple of terms that could be included in any attempt to discuss strategic planning or situational analysis. Generally, they are words that fit in because they are used interchangeably with other words that fit in as well. *Tactics* is one of these words. In many cases, tactics is the word used to describe the smaller steps necessary to achieve an *objective*, and it is used in the same way the term *objective* is used to describe the steps required to achieve or attain a goal.

Intentions

in·ten·tion (in-ten'shən) *noun* 1. A course of action that one intends to follow. 2. An aim that guides action; an objective.

The word *intention* can be used as a synonym for goal, purpose, end, aim, or objective. All of these commonly refer to what one hopes to *do* or *achieve*. The dictionary suggests that the word *intention* simply signifies a course of action that one *proposes*. Intention is not a particularly strong word in that it doesn't communicate a terribly deep sense of *purpose*. *Intent* implies something somehow more deliberate, and the other words synonymous with goal vary in intensity depending upon the context. Therefore, I'm not convinced that these terms are always interchangeable and that's where my problem begins.

A goal is something you hope to achieve or attain—something you have thought about and committed yourself to. An intention is something else entirely. Let me point out the difference by asking a question.

Let's say there are three frogs sitting on a log when one of the frogs decides to jump off, as seen in Figure 2-1. How many would remain?

Figure 2-1 Intention and deciding are two entirely different things!

If your answer was two, you're wrong. Even though it was the frog's earnest *intention* to jump, deciding to jump and actually jumping are entirely different things!

The key to success may very well lie in recognizing the steps involved in moving from the *intention* to jump to the physical act of jumping itself. However, the courage and strength required to take that great leap of faith may have as much to do with your perceptions of where you hope to land as it does with your discomfort and frustration with where you have currently been.

A SNAPSHOT OF WHERE YOU ARE

Work

work (wûrk) *noun Abbr.* wk. 1. Physical or mental effort or activity directed toward the production or accomplishment of something.

Work is what we do, not who we are. That's hard to believe in a culture like ours, but it's important to recognize nevertheless. Work is a place we go to dedicate our energy and effort in the pursuit of success—however we define it. But success isn't a measure of our worth or value as a human being. A *bad* day at work doesn't mean you're a *bad* person. Work is a means to an end, not the end itself, which is something we should always try to keep in perspective.

Most of us know someone who worked tirelessly to *give* her family everything she thought they wanted or needed, at the expense of just about everything else. Endless hours and lost weekends sacrificed at the *Altar of the Better Neighborhood*, the *Newer Car*, or the *Bigger Boat*. She worked hard to climb the ladder of success, only to find that those close to her—the very people she worked so hard for—were standing at the top of another building… a *different* building.

It is important to recognize that work is not the goal or even the objective. It is just one of the tools we use to achieve our goals or objectives. The easiest way to use work as a tool is to build a business that will work at least as hard for you as you work for it!

Income

in·come (in'kum) *noun* 1. *Abbr. inc.* The amount of money or its equivalent received during a period of time in exchange for labor or services, from the sale of goods or property, or as profit from financial investments.

I have always found it fascinating that most automotive service business owners have no idea how much money they actually make per hour, week, month, or year. They don't have a good idea of what they are being paid or how much they pay themselves. They don't have a good idea of the *perks* the business provides, or their value in terms of dollars and cents, either.

In too many instances, owners wind up subsidizing both their business and the repairs of their customers' cars and trucks by demanding far less compensation than they should be receiving, or are worth, for the time and effort invested.

To succeed, you have to meet or exceed your own wants, needs, and expectations *first*. Then, and only then, should you turn your attention to the wants, needs, and expectations of others. One way to do that is to inventory all the things you do for your business. Translate that into a job description and determine what you would have to pay someone else to do everything you are currently doing. If you aren't being compensated at least as well as you would have to pay someone else, you're cheating yourself, your family, and your business! Pay yourself first!

Determine what your wants, needs, and expectations are from the beginning. Create a realistic salary demand, use the reverse budget technique we will introduce later in this series to establish the required level of volume and profit, and then put the machine to work.

Freedom

free·dom (frē'dəm) *noun* 1. The condition of being free of restraints. 2. The capacity to exercise choice; free will: *We have the freedom to do as we please all afternoon.*

Freedom can be looked at in one of two ways when we discuss creating a snapshot of where we are in our professional lives. We can talk about the *freedom to* or we can discuss *freedom from*. Both are valid—both are real. The question in regard to your business is whether or not you give your business life through your sacrifice, or the business gives you— your family and everyone else who works with you—life by its efforts.

Family

fam·i·ly (fam'ə-lē, fam'lē) *noun Abbr.* fam. **1.** A fundamental social group in society typically consisting of a man and woman and their offspring. **2.** Two or more people who share goals and values, have long-term commitments to one another, and reside usually in the same dwelling place. **3.** All the members of a household under one roof.

If you think about it, each of us has many families. We have a *work* family and an *immediate* or *nuclear* family. We may even have a *social* or *recreational* family—groups of people we see when we are at play. In many cases, we spend more time with our work family than we do with our family at home, so having positive and powerful relationships with those who work with us is important.

It is equally important to maintain healthy relationships at home. It is critical to any discussion of success. You can't consider yourself successful if one of your values is family and you are away from them six or seven days a week, ten or twelve hours a day.

The result of that kind of conflict is frustration at the least—loss of your family and a complete mental and physical collapse at the worst.

Relationships

re·la·tion·ship (ri-lā'-shən-ship') *noun* **1.** The condition or fact of being related; connection or association. **2.** Connection by blood or marriage; kinship. **3.** A particular type of connection existing between people related to or having dealings with each other.

Your relationships tend to be defined by the roles you accept and agree to play. Family relationships center around the home—professional relationships flow from the business.

As you come to understand the various types of relationships, remember that most of us spend far more time at work than we do anywhere else. In fact, it is likely that you see the people at work more than twice as much as you see your spouse, children, or parents. This can be a serious problem for many people struggling in the automotive service industry. The solution our family ultimately discovered was interesting. Virtually all of us (all three sons) fell into the orbit of our father's automotive service business. At one time or another, all three of us worked with him in the same shop (two of us still do), and all of us still live within twenty minutes of each other.

Relationships with individuals that share the same principles and values, and have similar goals and objectives, tend to be enduring relationships. Relationships based on expediency or convenience usually are not.

Time

time (tīm) *noun* 1. *Abbr.* t., T. **a.** A nonspatial continuum in which events occur in apparently irreversible succession from the past through the present to the future. **b.** An interval separating two points on this continuum; a duration. **c.** A number, as of years, days, or minutes, representing such an interval. **d.** A similar number representing a specific point on this continuum, reckoned in hours and minutes. **e.** A system by which such intervals are measured or such numbers are reckoned.

Time is perhaps the most democratic of all phenomena in the universe: each of us has the same sixty seconds every minute, the same sixty minutes every hour. The richest man in the world could not buy an extra second no matter how much he wanted to spend, because time plays no favorites and knows no master.

You can't control it. You can't master it. You can't even manage it! All you can do is manage yourself within it. To be successful, you have to get the most out of every second you have been given.

Consequently, if you are ever to establish a picture of where you are, you will have to do that in the context of the time you have and are willing to devote to your business life. It is better to establish that at the beginning of the planning process than halfway through it or at the end. Your business can be the worst possible kind of mistress, demanding everything and offering nothing in return. Or it can be like the best of all partnerships, providing you the support and assistance you need to succeed.

PRINCIPLES & VALUES

In moving from intent to implementation, we have to identify the principles that govern our lives and the values that are important to us. We have to realize that the consequence of ignoring those values or compromising those principles is conflict, frustration, and ultimately anger or even self-loathing. We have to establish our roles and define our responsibilities. Until we do that, we cannot effectively identify our objectives or establish appropriate goals.

We must understand where we are in our lives as well, especially with regard to work and our relationship to it. We have to ascertain our wants, needs, and expectations, and we have to reconcile them with our own salary demands (income), freedom, family, relationships, and time as we develop a set of objectives for our work-centered lives.

Someone once said that the greatest journey starts with the first step. I'm not one hundred percent sure that's true. In order to plan any journey—before taking your first step—you must know two things: where you are and where you are going. I think you will agree that we've just laid the groundwork for accomplishing both.

Once we know who we are, what is important to us, and what our principles are, we can begin to explore where we want to go without having to worry if we find ourselves lost along the way. Understanding where we are with regard to work, income, family, time, and relationships helps us understand—or at least have a better sense of—where we are going. It acts as a moral compass to help keep us on track, and it allows us to recognize when we stray off course.

SUMMARY

Creating a personal inventory of values, goals, and objectives can be one of the most difficult things any of us will ever undertake. It forces us to look deep into our own souls for the answers to time's most uncomfortable questions, such as: What is really important to us? Creating an inventory of business goals and aspirations can be equally difficult because it forces us to realize that anything we do that is not consistent with our principles can only result in conflict and frustration.

Write down some of your personal values, goals, and objectives, along with the principles that you feel guide your life in the chart on the next page. Accompany them with a list of things you hope to accomplish in business.

Thoughts

Actions

Results

CHAPTER 3

Where Do You Want To Start?

INTRODUCTION

This chapter will introduce you to the concept of benchmarking—that is, creating a baseline of performance for your business. It will also define the ten critical Key Performance Indicators (KPIs) necessary to effectively monitor and manage your automotive service business:

- Gross profit margin on labor
- Gross profit margin on parts
- Car count
- Average invoice (or average repair order)
- Labor mix
- Labor content per job
- Technician efficiency
- Service bay productivity
- Total sales per bay
- Total sales per technician

You will also learn how to use these critical KPIs to move your business toward success.

CREATING A BASELINE

A journey like this begins by firmly establishing where you are in your business at a particular moment in time. Knowing where you want to end up is irrelevant if you have no idea where you will start the journey. You figure that out by identifying the processes in your business that are critical to success and then breaking those processes down into metrics that will allow you to accurately measure performance. The results of such an exercise can be called many things, but I call them Key Performance Indicators (KPIs).

If you understand what the key processes of your business are and know what great performance (the targets you are shooting for with regard to those numbers) looks like, you can quickly identify just how well (or not so well) you are doing.

KPIs: Benchmarking

There is probably no limit to the number of things you can measure in an automotive business, ranging from basic productivity to the cost per square foot of service bay floor space. Someone could write a book on this one subject alone, but I'm not sure that it would help as much as achieving a basic understanding of the principles we will discuss in this section and applying this discussion to the real world. Consequently, we will only be looking at ten of these KPIs. However, the ten we look at will provide nearly all the information we need to determine the physical health of our business at any given moment, and certainly what our business looks like when compared to another business in the same industry.

The ten KPIs we are going to look at are Gross Profit Margin (GPM) on labor, GPM on parts, average invoice, car count, labor mix, labor content per job, labor sales or total sales per bay, labor sales or total sales per tech, efficiency, and productivity.

Just a note about what we are *not* going to look at. We are *not* going to look at net profit for an automotive service business because too many external influences can alter that number dramatically—such as whether the business is a sole proprietorship (owned by a single entity), a partnership, or a corporation. We are not even going to consider gross profit on the entire business, other than to try to understand how it is impacted by these other ten KPIs because *how* and *where* you report your financial information on the income statement can have a significant impact on it as well. We will, however, spend a fair amount of time covering these ten KPIs because without knowing what they are and what they tell you about your business you are piloting an airplane that has no windows!

We have said repeatedly throughout this series that if you can't measure it, you can't manage it! Now we're going to take a look at what needs measuring.

GPM on Labor

Generally speaking, gross profit on labor can be defined as the hourly labor rate less the cost of direct labor (the technician's wages or salary per hour associated with making sales). GPM on labor can be expressed by dividing gross profit on labor by the hourly labor rate with the result expressed as a percentage. Gross profit on labor can be reported *loaded*—reflecting benefits and ancillary costs like health insurance, sick or personal days, vacations, uniforms, and other related expenses—or it can be reported without the load (*unloaded*). My suggestion would be to move all the direct costs associated with producing labor sales to the top of your financial statement (covered in another book in this series) where you or your bookkeeper or accountant generally reports cost of goods sold (the cost of the parts and accessories you purchase for resale).

The cost of direct labor would include all the costs we just mentioned—sick or personal days, uniforms, vacation pay, holiday pay, as well as life, health, and disability insurance. This will give you a far more accurate picture of what your overall gross profit and GPM are because it will more accurately represent the costs involved in producing those labor sales.

Before you get confused, the difference between gross sales (total sales) and those sales minus the costs associated with those sales (sales minus the cost of sales = profits) can be reported in two ways. Gross profit is almost always presented by your bookkeeper or accountant in *dollars*, while GPMs are almost always presented in *percentages*.

A formula reflecting gross profit on labor would look something like this:

Hourly Labor Rate – Cost of Direct Labor per Hour = Gross Profit on Labor per Hour

Let's give this formula life by substituting descriptions with actual numbers. Let's say your labor rate per hour is $65.00. You pay your technician an hourly rate of $17.20, unloaded. The load of taxes, benefits, and associated expenses can drive your total direct cost of labor up anywhere from twenty to thirty percent from the technician's hourly unloaded rate. Splitting the difference at a twenty-five percent load would add $4.30 in associated costs to the technician's hourly rate—a total direct labor cost of $21.50.

Knowing your hourly labor rate and your cost of direct labor allows you to subtract one from the other—$21.50 from $65.00 leaves you a gross profit of $43.50 and a GPM of 67 percent.

As a general rule of thumb, your GPM on labor should range somewhere near 60 percent with the benefits and taxes loaded in and about seventy percent with them removed.

Rule of Thumb

Gross profit on labor: approximately 70 percent with core benefits, 60 percent without.

Gross profit calculation

$65.00 (labor rate) – $21.50 (labor cost per hour) = $43.50 (gross profit on labor)

GPM calculation

Divide gross profit by labor rate

Gross Profit ÷ Hourly Labor Rate = GPM (expressed as a percentage)

$43.50 ÷ $65.00 = 66.9%

Your numbers

$_____._____ ÷ $_____._____ = _____ %

GPM on Parts

You calculate gross profit and GPM on the parts and accessories you sell using the same principles you use when calculating gross profit and GPM on labor. To do this, start with the parts price (the price the customer will pay for that part or accessory), then subtract the cost of goods sold or the *acquisition cost* (the cost for you to acquire that part). The result is your gross profit on parts.

Stated another way, the formula would look like:

Selling Price (parts price) – Acquisition Cost (cost of goods sold) = Gross Profit on Parts

For example, a part that will be sold for $60.00 cost you $36.00 to acquire from your local parts supplier. We can give this formula life by once again substituting actual numbers into the equation. To calculate the gross profit, you subtract the cost of the goods you are selling from the selling price and what is left is your gross profit: $60.00 – $36.00 = $24.00.

Rule of Thumb

Gross profit on parts should average 45 to 55 percent

Gross profit calculation

$60.00 (selling price) – $36.00 (acquisition cost) = $24.00 (Gross profit on parts)

GPM calculation

Divide gross profit by selling price

Gross Profit ÷ Selling Price = GPM (expressed as a percentage)

$24.00 ÷ $60.00 = 40.0%

Your numbers

$_____._____ ÷ $_____._____ = _____ %

Knowing the gross profit (represented in dollars) will allow you to calculate your GPM on parts, which is also a critical number for you to know and understand. To calculate the GPM on parts, you divide your gross profit on parts by the selling price of that part or accessory: $24.00 ÷ $60.00 = 40 percent.

I used a GPM on parts of 40 percent here, not because it is the target but because it is probably the minimum average GPM on parts you can accept and still hope to remain in business. Overall, GPM on parts should average 45 to 55 percent. Low margins will compromise your ability to make a decent living, invest in your business, and possibly even pay your bills. However, higher margins can often affect sales in a negative way by discouraging the more price sensitive segment of your market from trading with you. For this reason, a number of automotive repair shops utilize a concept known as matrix pricing for their parts sales, which utilizes higher margins for the lower-priced parts and lower margins for the higher-priced parts.

A parts pricing matrix for items in your inventory, or for parts acquired from a local supplier (which are generally purchased at a deeper discount than parts purchased from the local dealership), could resemble that seen in Table 3-1.

Table 3-1 Normally Sourced Parts Matrix

Parts Cost	Desired GPM	Divide By
$0.00 – $5.00	70.0%	0.30
$5.01 – $10.00	60.0%	0.40
$10.01 – $75.00	55.0%	0.45
$75.01 – $200.00	50.0%	0.50
$200.01 – $500.00	42.5%	0.58
$500.01 – $750.00	35.0%	0.65
$750.01 and up	30.0%	0.70

The parts pricing matrix for dealer-sourced parts could resemble that seen in Table 3-2.

Table 3-2 Dealer Sourced Parts Matrix

Parts Cost	Desired GPM	Divide By
\$0.00 – \$1.00	70.0%	0.30
\$1.01 – \$5.00	65.0%	0.40
\$5.01 – \$25.00	60.0%	0.45
\$25.01 – \$75.00	55.0%	0.50
\$75.01 – \$500.00	42.5%	0.58
\$500.01 – \$750.00	32.0%	0.65
\$750.01 and up	28.0%	0.70

The overall philosophy is simple. Parts that fall below a certain dollar threshold aren't the parts to which the consumer is normally sensitive. The vast majority of the sales that make up your parts mix fall within the category of prices under \$100.00. Most consumers are far more sensitive to the price of a \$2,300 rebuilt transmission and the profit you are taking on it than they are to the nuts and bolts that cost them \$.95 but only cost you \$.20.

Regardless of how you calculate GPM on parts—whether you use the matrix models or the formulas we've just discussed—you must remember that adequate parts margins are the key to success in any business.

Car Count

Car count is another critical number when it comes to benchmarking your business. Your car count is simply the number of cars that pass through the shop every day, week, month, or year. It includes *all* of the cars that come to the shop, even the vehicles for which there is no charge. If you use sequentially numbered invoices, subtract the lower number (the one that appears on the first invoice for that period) from the larger number (the number that appears on the last invoice or repair order for that period). Then, divide that number by the number indicating the appropriate time period.

Car Count = Total Vehicles ÷ Calendar Period

For example, if the first invoice for the month was number 23390, and the last invoice for that month was number 23588, you would subtract 23390 from 23588 and the total number of invoices—or the *car count*—for that month would be 198. If you were calculating daily car count and that particular month had 22 working days, you would divide 198 by 22 for a daily car count of nine cars per day, as seen in Table 3-3.

Table 3-3 Car Count

Start Date	July 1, 2002	Invoice Number	23588*
End Date	July 31, 2002	Invoice Number	23390
		Total Invoices	198
		Divided by Working Days	22**
		Equals Daily Car Count	9

* All invoices, including $0.00 invoices ** Reflects July 4 Holiday

Car count is important because of the impact it has on scheduling, productivity, and technician efficiency, among other things. When the car count is too high, or at least high enough to become overwhelming for the number of technicians available, efficiency and productivity begin to erode. When there are too many cars being forced into the pipeline, the quality of the work performed on the cars coming out of the pipeline will more than likely be compromised. Quality will suffer as accuracy is sacrificed at the *altar of expediency*.

Higher car counts generally equal lower productivity and lots of undone service work identified and itemized in the *remarks* or *recommended service* area of the invoice. These lost service opportunities are synonymous with lost profits in a world in which time is one of the most precious commodities any of us may have. Getting the vehicle back to the shop may be far more difficult than getting the work done while the vehicle is already in the shop.

So, the key to *car count* isn't just working on fewer cars—the key is working on fewer cars, but doing more work on the cars you are working on.

Average Invoice (Average Repair Order)

Of all the KPIs, *average invoice* is perhaps the most telling. Although it may be the simplest to understand and the easiest to calculate, it can furnish more information about an automotive service business—particularly an independent automotive service business—more than any other single KPI.

Average Invoice = Gross Sales ÷ Car Count

Your average invoice is the product of your gross sales for any given period of time divided by your car count. If we were looking at a three-bay service facility with average monthly gross sales of $33,333 (a $400,000 per year enterprise) and a monthly car count of 198 vehicles, the average invoice would be calculated by dividing $33,333 monthly sales figure by the vehicle count of 198 per month. In this case, our average invoice would be $168.35 (very close to the national average), as seen in Table 3-4.

Table 3-4 Average Invoice

Average Monthly Sales	$33,333
Divided by Car Count	198
Equals Average Invoice	$168.35

High average invoice usually correlates to lower car counts and higher profits in most shops. The higher the average invoice, the higher the productivity—the higher the productivity, the higher the profits. High productivity and higher profits are the product of thoroughly inspecting the vehicle, analyzing service needs, presenting those needs to the vehicle owner, and asking for the opportunity to perform the service. That translates into more service work per vehicle and results in a technician performing multiple service operations on the same vehicle.

Lower average invoices generally mean higher car counts, and higher car counts almost always indicate a high rate of *turn and burn*. A shop in which cars are blowing in and out of the service bays with no time for thorough inspection, service, or repair is a victim of this scenario. Vehicles are leaving the facility with undone service work that just might find its way out of your operation and to the door of another service provider who understands just how difficult it is for most vehicle owners today to get a vehicle into the shop in the first place.

Labor Mix

Labor mix is the percentage of labor sales to total sales as compared to the percentage of parts sales to total sales—expressed as a ratio. Sounds kind of confusing, doesn't it? But, it isn't really.

Until very recently, automotive shop owners (particularly independent automotive shop owners) were taught that they should work toward $1.00 worth of parts sales for every $1.00 worth of labor sales. In other words, total sales should be split 50/50 between the sales of parts and labor achieving a ratio of 1:1.

Once upon a time in this industry, that may have been *good enough*, but it isn't any more. With automotive service work becoming continually more and more labor intensive, a 50/50 or 1:1 model just won't work any longer. There are too many labor dollars evaporating off the service bay floor.

Looking at that model, we can see illustrations of a number of the KPIs and how they work together to paint a picture of Average Automotive, Inc.—a company we have created to serve as a composite example of everything we have been talking about to this point.

Looking at one hour's worth of work consistent with the old paradigm of a 1:1 or 50/50 parts/labor split ($60 parts/$60 labor) as illustrated in Table 3-5, the following numbers jump out at us—or at least they should. Our direct labor cost has increased to $22.93 due to an increase in health care costs. That number reflects the load of taxes and benefits. Our gross profit on labor is within the 60 percent GPM we're looking for with the taxes and benefits *loaded* in (70 percent without). Our gross profit on parts is sitting at 40 percent, which is the minimum we'd like to see as an average GPM for parts sales. And, when we look at *profit per hour,* we received $120.00 for one hour's worth of work: $24.00 gross profit on parts and $37.07 gross profit on labor, with a total gross profit of $61.07 (or just about 51 percent).

Table 3-5 50 Percent Labor – 50 Percent Parts

Labor Sales	$60.00
Gross Profit Margin	62%
Gross Profit	$37.07
Cost of Labor	$22.93
Parts Sales	$60.00
Gross Profit Margin	40%
Gross Profit	$24.00
Cost of Parts	$36.00
Total Sale	$120.00
Total Gross Profit	**$61.07**
Gross Profit Margin	50.9%

All in all, that doesn't seem quite so bad and it isn't compared to the extremely poor performance of a majority of shops engaged in automotive service and repair. However, it's not as good as the performance we'd like to see—performance that starts to improve the minute we tip the scale and begin to recover some of the lost labor dollars slipping through our fingers.

These labor losses are the result of a shift in the industry from low-tech or no-tech, to high-tech and high-touch—from removal and replacement (R&R) to inspection, testing, analysis, evaluation, diagnosis, and repair. Being able to ascertain what the problem might be and, perhaps more importantly, what might be causing it has become an increasingly more important part of basic shop operations as vehicle sophistication and technology have blossomed. Determining what needs to be done has become just as important as knowing how to perform the appropriate repair. Vehicles require two and three road tests to identify symptoms and then confirm proper correction and all of this takes time that can be translated into labor hours that aren't always being charged out on the customer's repair order.

To be successful in the automotive service environment today requires a parts/labor split that more closely approaches 60/40 than it does 50/50, with labor sales moving as close to 60 percent of total sales as possible. When this begins to happen, the results are almost instantaneous and very dramatic, as you can see in Table 3-6.

Table 3-6 55 Percent Labor – 45 Percent Parts

Labor Sales	$74.00
Gross Profit Margin	69%
Gross Profit	$51.07
Cost of Labor	$22.93
Parts Sales	$60.00
Gross Profit Margin	40%
Gross Profit	$24.00
Cost of Parts	$36.00
Total Sale	$134.00
Total Gross Profit	**$75.07**
Gross Profit Margin	56.0%
"New" Model Gross Profit	$75.07
"Old" Model Gross Profit	$61.07
Difference	$14.00

You might notice a couple of differences between the *old* model and the *new* model almost immediately. First of all, you *should* have noticed that the labor rate increased from $60.00 for one hour's worth of work to $74.00. It had to. No one wants you to sell fewer parts in the same hour's worth of work. We would, however, like to see you recover some of the lost labor time we've been talking about, and the easiest way to do that is to charge more per hour for every hour's worth of work you charge for! If our parts sales remain the same, and that number becomes anything less than 50 percent of total sales, our percentage of labor sales for the same period of time must go up. Consequently, the increase is reflected above. Our direct cost of labor remains the same for our purposes here, although you will see in just a moment that we could conceivably invest more money in direct labor, perhaps resulting in the employment of a somewhat better technician.

How or why? Because with all things remaining virtually the same with the one exception of an increase in labor sales, our gross profit jumped to $75.07 or $14.00 more than the same hour's worth of work resulted in using the old model.

There are more advantages to embracing this kind of paradigm than can be discussed here; however, don't be fooled. The advantages we look at here can and will make a powerful and positive difference in your business. First of all, our *profit per hour* increased by $14.00, that is almost a 23 percent *bump* in gross profit for the same hour's worth of investment. If you can invoice eight hours in an eight-hour day and sell the appropriate amount of parts, you could realize an increase in revenue of $112 per day, and that is with a 55/45 split, not the 60/40 split that is the target.

This bump occurs because the GPM on labor is substantially higher than the overall GPM on parts and any increase in labor sales is reflected on the bottom line almost instantly.

You really don't have to calculate this ratio on your own. The numbers already exist, and if you have a computerized automotive shop management system, a bookkeeper, or an accountant, finding them is really pretty straightforward. In most financial reporting systems each component on the income statement is reported as a percentage of total sales. Consequently, all you need to do is look at the statement, which looks very much like the model statement for Average Automotive, Inc. you see in Table 3-7 and you will begin to see the correlation we are looking for.

Table 3-7 Average Automotive, Inc. Statement of Revenue and Expense, Period Ended August 31, 1999

	Aug. 31, 1999		Aug. 31, 1998	
	Amount	Percent	Amount	Percent
Revenue				
Sales – Labor	$46,292.16	55.04%	$41,641.41	55.33%
Sales – Parts	$37,806.89	44.96%	$33,622.84	44.67%
Total Revenue	$84.099.05	100.00%	$75,264.25	100.00%

As you can see in Table 3-7, labor sales and parts sales at Average Automotive have reflected a consistent 55/45 split for the two-year period reported, and sales have increased by $8,834.80. Sales have increased by almost 12 percent while the percentage of parts and labor sales has remained constant, which should mean a significant increase in profits.

Labor Content Per Job

Labor content per job is one KPI that is often overlooked, yet it can play a critical role in solving or at least ameliorating one of the most common complaints in the industry—the inability to efficiently schedule work in the shop. I stumbled across using *labor content per job calculations* to help better regulate the flow of work in our shop almost by accident while in the process of establishing a set of KPIs for our business. I was working on average invoice calculations when I began to wonder what role (if any) labor mix might play. After all, if the labor mix percentages are accurate (and they have to be), and your average invoice is actuarially sound, you should be able to apply the parts/labor ratios to your average invoice.

What do you gain by knowing what the *average parts dollars per job* or the *average labor per job per average invoice* is? I wondered the same thing… until I actually did the math. Then, it hit me! If you know what the average labor dollars per job are and you know what your labor rate is (preferably, the effective labor rate that we will be discussing below) you can divide average labor dollars per job by your labor rate. The result will be your labor content per job, as seen in Table 3-8.

Table 3-8 Labor Content Per Job

Average Invoice		$168.35	
Parts Sales in %	45%	$75.76	Parts Dollars Per Job
Labor Sales in %	55%	$92.59	Labor Dollars Per Job
Labor Dollars Per Job		$92.59	
Divided by Labor Rate		$65.00	
Labor Content Per Job		1.4	

The numbers are telling us that we can expect to spend an average of 1.4 hours on every vehicle that pulls down the driveway. Obviously, there won't be an hour and a half spent on every car... but, the average is actuarially sound because all the numbers that led us to it (car count, average invoice, labor mix, etc.) are actuarially sound.

Consequently, we can schedule about 5.3 cars for each technician capable of completing eight hours worth of work in an eight-hour day or close to 100 percent productivity. If we have ten cars on the lot, we know there will be 14 hours of work to be done regardless of what the vehicles might have been brought to the shop for.

In a minute we will be looking at *service bay productivity*. One of the concepts that will flow out of that discussion will be *effective labor rate*. Effective labor rate is another critical concept and when we get there we will revisit the same calculation we just developed to see first-hand what a dramatic difference productivity can make in your business.

Technician Efficiency

ef·fi·cien·cy (e-fish'ən-sē) *noun Abbr.* eff. **1. a.** The quality or property of being efficient. **b.** The degree to which this quality is exercised: *The program was implemented with great efficiency and speed.* **2. a.** The ratio of the effective or useful output to the total input in any system. **b.** The ratio of the energy delivered by a machine to the energy supplied for its operation. *An efficient person is competent or productive to a high degree upon some actual or imagined scale or model.*

Generally, the issue of technician efficiency doesn't come up until after the discussion of service bay productivity has been addressed. This time, I'd like to do things a little differently and begin with technician efficiency instead of ending with it.

We will define *technician efficiency* as the relationship between the time that has been sold or allotted for the completion of a specific service operation and the time it actually takes the technician to perform that operation.

Stated another way,

Hours Sold ÷ Actual Time = Technician Efficiency

The *hours sold* are the same as *hours invoiced*: they are the number of hours you multiply with your labor rate to determine how much you will charge the customer for the labor to perform the service, replace the part, or complete the maintenance operation. *Actual time* is the time it took to complete the job, which is rarely the same high or low, as hours sold.

If the service writer at Average Automotive invoiced the customer $65.00 to R&R an alternator (which correlates to Average Automotive's $65.00 per hour labor rate), and the R&R actually took the technician the hour that was invoiced, technician efficiency for that particular job was 100 percent. If, on the other hand, it took the technician an hour and a half to remove and replace the alternator, technician efficiency would drop to 67 percent. With the proper tools, the proper training, the proper equipment, and the opportunity to succeed—such as the on-time arrival of the proper alternator—it is possible that the technician might have been able to remove and replace the alternator in less than one hour. In this case, technician efficiency would be *greater than* 100 percent, as seen in Table 3-9.

Table 3-9 Technician Efficiency

Hours Sold	+	Actual Time	=	Technician Efficiency
50	+	40	=	125%
50	+	50	=	100%
40	+	50	=	80%

The *responsibility* for technician efficiency falls at the feet of the technician, provided he or she is afforded the proper tools, training, technology, and opportunity to get the job done. Productivity, as we will see in a minute, is a *front counter* management responsibility.

Service Bay Productivity

We will define *service bay productivity* as the relationship between the hours billed (or labor hours invoiced) and the labor hours available for sale (the hours the technicians are on-site and available to work).

Hours Billed ÷ Hours Available to Work = Productivity

Productivity has as much to do with the ability to sell the hours you have available for sale as it does actually getting the work done. In fact, it may actually have more to do with selling those hours than anything else!

High productivity = high profits. Low productivity = low profits. It doesn't get any simpler than that.

If you have one technician and that technician works 40 hours per week and you only sell 36 of those 40 available hours, you divide 36 by 40 and service bay productivity would equal 90 percent. If you sell 40 hours and the technician is able to complete the 40 hours of work you sold in the 40 hours they were there, service bay productivity would be 100 percent. And, finally, if you sold 50 hours of labor that your technician was then able to complete in the 40 hours they were on-site and working, service bay productivity would soar to 125 percent, as Table 3-10 points out.

Table 3-10 Service Bay Productivity

Hours Invoiced	+	Hours Avail.	=	Productivity
36	+	40	=	90%
40	+	40	=	100%
50	+	40	=	125%

Unfortunately, statistics suggest that service bay productivity numbers leave a lot to be desired, especially for the independent segment of the service industry. And the tragedy there is the profound impact productivity has on every aspect of a service business. The above example is representative of one technician. But what happens when there are three technicians working in the shop? They won't share the same numbers, especially if they are segmented—and by that I mean specializing in a certain area of repair. For instance, you might have one technician doing high-tech, drivability diagnosis and repair—focusing on intermittent and erratic electronic problems. Another technician might be doing nothing but brake and front-end work, and still another might spend most of his or her time focusing on light and heavy line. Their productivity number will reveal a significant *spread* in most shops around the country, with the highest productivity numbers coming from the brake and front-end specialist, the lowest numbers coming from the drivability specialist, and the hours closest to 1:1 (one hour billed for every hour worked) coming from the technician doing the light and heavy line work.

Average service bay productivity of 87 percent would actually be something you could be proud of, especially when most of your peers are posting numbers below 70 percent, as seen in Table 3-11. But that isn't the issue I would like to focus on for the moment. I would like to turn your attention to what the numbers mean to you and possibly your shop.

Table 3-11 Average Productivity (3-Technician Shop)

	Sold Hours	Available Hours	Productivity
Driveability Tech	26	40	65%
Line Tech	34	40	85%
Brake & Front End	44	40	110%
Total	104	120	87%

Service bay productivity directly impacts a KPI I refer to as *effective labor rate*, or the labor rate you have posted on the wall factored by service bay productivity. What does that mean? Let's work with the same $65.00 per hour labor rate Average Automotive has been posting through most of this discussion. In most cases, that's the labor rate Average Auto's owner *thinks* he's getting for every hour he bills. It isn't! Because his shop is running at an average service bay productivity of 87 percent, he is actually only receiving 87 percent of $65.00, or $56.55.

Labor Rate x Average Service Bay Productivity = Effective Labor Rate

If Average Automotive was realizing an effective labor rate closer to the averages I've seen over the years—the numbers reflected in most productivity surveys—the effective labor rate would be closer to $35.75. In another book in this series, the book that deals with achieving *operational excel-*

lence, we'll take a long, hard look at all the factors that can impact and influence service bay productivity. But for now, I'd like to turn the discussion to effective labor rate and labor content per job.

When we looked at labor content per job earlier in this chapter, we calculated that number by taking the dollar amount of our average invoice and multiplying it times the percentage of labor mix found in our financial statement or calculated through our computerized automotive shop management software. The result was the labor content per job expressed in dollars. We then took that number and divided it by our labor rate to give us our labor content per job expressed in hours. But, we just saw that would only—*could only*—be accurate if our productivity was 100 percent! In most shops across the country, we're not receiving one hour's worth of revenue for every hour we're there because we're either not selling every hour we have to sell or our technicians are unable to produce at 100 percent technician efficiency. Consequently, in order for our labor content per job (expressed in hours) to be accurate, it must be factored for service bay productivity!

Look at Table 3-12 to see what happens when we do that.

Table 3-12 Labor Content Per Job

Average Invoice		$168.35	
Parts Sales in %	45%	$75.76	Parts Dollars Per Job
Labor Sales in %	55%	$92.59	Labor Dollars Per Job
Labor Dollars Per Job		$92.59	
Divided by Effective Labor Rate		$56.55	
($65.00 x 87% = $56.55)			
Labor Content Per Job		1.6	

Our labor content per job went from 1.4 hours to 1.6 hours. But what does that really mean? It means that it will take us two tenths of an hour longer on each car we see in the shop—or one less car per day… $168.35 less per day in revenue.

If we modify the labor content per job to bring it more in line with national averages—adjusting the effective labor rate to reflect average productivity of 55 percent—the numbers get even worse, as seen in Table 3-13.

Table 3-13 Labor Content Per Job

Average Invoice		$168.35	
Parts Sales in %	45%	$75.76	Parts Dollars Per Job
Labor Sales in %	55%	$92.59	Labor Dollars Per Job
Labor Dollars Per Job		$92.59	
Divided by Effective Labor Rate		$35.75	
($65.00 x 55% = $35.75)			
Labor Content Per Job		2.6	

The labor content per job goes up, but the size of the average invoice does not! It now takes us 2.6 hours to complete $92.59 worth of labor billing, and we can only effectively work on three cars a day.

High productivity is critical to your success as a shop owner or manager. It impacts everything from bottom line profits to scheduling—from staffing to just about everything else that happens in the shop. When we begin to discuss operational excellence, we'll take this critical KPI apart, and then we'll take it apart again!

Total Sales per Bay

The final two KPIs are *total sales per bay* and *total sales per tech*. Total sales per bay is relatively simple to calculate. You just take the total sales for the period you are focusing on and divide it by the number of bays you have. If Average Automotive had three bays and sales of $33,333 per month, total sales per day for that month would be $11,111.

That is the actual number. But, what is the target?

The number I have heard thrown around most often is $20,000 to $25,000 per bay per month, and it is calculated in Table 3-14's projection—assuming one technician in each service bay.

Table 3-14 Total Sales Per Bay

Techs Per Bay	1
Hours Per Day	8
Days Per Month	22
Tech Hours Per Month	176
x Labor Rate	$65.00
Total Possible Labor $	$11,440.00
Plus Est. Parts Sales	$11,440.00
Total Sales Per Bay	**$22,880.00**

Unfortunately, the number can be misleading. Not every shop has one technician for each service bay, and consequently—since the technician is the revenue generator, not the service bay—the calculations fall apart. Aside from that, few shops run at 100 percent productivity (8 hours of billing per day, which is what the calculation above would suggest). In fact, looking at the actual performance of Average Automotive versus the target projection, what do you see? Actual performance of about half the projection—a target of $68,640 (3 techs x $22,880) and actual sales of $33,333—or just under 50 percent productivity.

In this case, a more accurate KPI might be *total sales per tech*.

Total Sales per Tech

Total sales per tech are calculated the same way as total sales per bay are calculated with one exception—the number of service bays is irrelevant. However, one thing might become obvious if you play with the same numbers we used above. Average Automotive has three technicians and total sales of $33,333 per month. Our calculations would suggest that if Average

Automotive were working at 100 percent productivity, those sales should be closer to $100,000 per month—not one-third of that!

SUMMARY

There are a number of critical elements involved in any successful journey. One is a deep understanding of where you are. Another is knowing where you would like to be. A third is a realistic appraisal of just how you are going to get there and a fourth involves recognizing when you are finally there!

In business, one of the ways you accomplish that is by base-lining—creating a snapshot of exactly where your business is when you start this journey toward success. You do that by identifying the critical business metrics that measure performance in your industry. These are generally referred to as KPIs.

The key to KPIs is using what those numbers tell us to drive positive performance. The numbers are only numbers if we don't use them to create and sustain total quality and continuous improvement.

The key to success in our industry is to raise *productivity*, increase *average invoice*, develop a higher percentage of *labor sales to total sales* while maintaining or increasing *parts sales* at the same time and growing *labor content per job*. If we can do all that, we will begin to see our bottom line begin to float up to where it really belongs!

Identify the KPIs you feel are critical to your business and write them in the chart on the next page.

Thoughts

Actions

Results

CHAPTER 4

Vision, Mission, Goals and Objectives

INTRODUCTION

In this chapter we will learn how to:

- Define our vision
- Understand our mission
- Recognize our values & guiding principles
- Establish personal & professional goals & objectives
- Utilize a technology for setting and achieving goals
- Manage process, analysis, and application

MISSION, VISION, GOALS, AND OBJECTIVES

Once you have taken a snapshot of your business by base-lining the KPIs we have been talking about, you have completed the first step in the planning process—establishing where you are. At that point, you are almost ready to start the journey—almost ready to begin moving from *intent* to *implementation.*

Like creating a *picture* of your business by capturing the numbers in the form of KPIs, the next step looks deceivingly simple. After all, once we know where we are, all we need do is figure out where we want to be. It's really no different than planning a vacation or any other kind of trip. How hard could it be?

The truth of the matter is it's harder than it looks and we should never lose sight of the fact that there are some people who never plan anything, let alone something as complex as a vacation or a long journey. They just get in the car, truck, or van and go!

Moving your business from where it is to where it needs to be isn't as easy as it may sound. Imagining your company as it *should* be is one thing—transforming that vision into reality is quite another. According to George W. Pratt III, Director of the Harvard Business School Publishing Corporation, it takes "a myriad of forces: political—*building a coalition behind your initiative*, emotional—*harvesting employees' ideas and gaining their commitment*, and logistical—*securing resources and managerial attention*." You can't do that without creating a compelling vision of an attractive, all-inclusive future—in other words, a vision that includes everyone you would like with you when you start your journey.

Roles Revisited

Before we turn our attention to personal vision, I'd like to revisit the concept of *roles* —discussed earlier in Chapter 2. In order for any of this to work, we must recognize that each of us functions in a number of different roles in the course of our lives. We are sons and daughters, fathers and mothers, and brothers and sisters just as much as we are employers and employees, owners and managers, and coaches and mentors. As we suggested earlier, each of these roles brings with it different responsibilities and expectations. To deny them removes a number of the critical factors that impact happiness profoundly.

If family is important to you, it is one of your personal values. If you find yourself in the role of father and husband as well as the role of business owner, you will have to find a way to reconcile the time your family is entitled to expect from you with the time running a business will demand from you. If you don't, the only possible result is frustration, anger, resentment, and disappointment. Consequently, you will have to establish a set of *ground rules* early in the process of developing both your personal and corporate visions that will help determine what your priorities are.

If your role as a *successful* businessperson takes precedence over your role as a parent or a spouse, the demands of family will prove a bothersome distraction at the least—a major annoyance and frustration at the worst.

There are more than enough *have to's* ready and willing to steal time away from your *want to's*, leaving you wondering where your life has gone, what you're working for, and why you're working so hard! Consequently, you must balance your roles and responsibilities with your vision and your mission. And while it is acceptable—even normal—to find these roles out of balance from time to time, balance and moderation are critical over the course of your lifetime for both long-term happiness and success.

Without understanding the many different roles we accept in the course of living our lives, establishing and achieving a personal or related vision becomes difficult, if not impossible.

Personal Vision

Developing a *personal* vision of the future comes first. Creating a *corporate* vision without first taking a look at your personal wants, needs, and expectations is like drawing a map to a city you may not want to visit.

Personal vision comes first for another reason—a more philosophical reason. Your business enterprise should be a tool that is used to make your personal dreams and aspirations possible—an extension of your personal vision. Attempting to create a corporate vision without first spending some time thinking about what you want this business to do *for* you is like leaving for vacation with no particular destination in mind—you never know where you are going, so there is no way to know if or when you get there.

If we're talking about vacations that may not be the worst thing that could ever happen to you. But, as far as your business is concerned, it's suicide. The more you allow your business to determine your direction, the less likely you are to end up where you want or need to be for you.

There is a technology for almost everything, and establishing personal vision is no exception. To paraphrase Stephen Covey you must *start with the end in mind.* That means having a clear mental picture of you functioning in the world as you would like that world to be. You can start that process by asking questions like the following:

> What do I want out of life professionally—what do I want out of life personally? What is important to me: family, material success, or service—the act of serving others? Is it possible for me to have it all?

Don't leave out any of the important areas of your world, like your life at home, health, work, personal development, community service, or material possessions.

What would your life look like if you had already achieved and acquired everything you ever wanted or needed? Do you think you would still want it after you got it? If you aren't happy with the *stuff* you have now, what makes you think more stuff will make you happier?

Once you have both asked and answered these important questions, you can ask yourself if the answers clearly define what it is you are really looking for out of life. If they do, they are probably worth pursuing and the only thing remaining is to see what stands in the way of achieving your vision. If they don't define what you are looking for, it may be time to revisit either the questions or the answers.

Remember, vision is important. Clearly defined, it determines what your personal mission will look like, what your personal goals will be, and ultimately what your corporate vision will encompass.

Personal Mission

Your personal mission is to pursue and achieve your vision. Your personal mission should be defined (and written down) before you start thinking about goals, objectives, strategies, or tactics. Too many people start the process backwards, constructing goals, tactics and objectives that aren't associated with the achievement of any particular mission or vision. Don't misunderstand—it's probably better to have goals and objectives that are not associated with a specific vision or mission than it is to have none at all. But, achieving those goals and objectives will be far easier if they are tied to a compelling and attractive vision of the future that is meaningful to you. It will be far easier to sacrifice your time and your energy when you can see and feel the future through your vision, mission, goals, and objectives.

One of the key components of my personal mission statement is:

> To benefit, enrich, and improve the lives of others by sharing knowledge and experience through my writing, speaking, and interactions with others in a very personal way.

It is one of the reasons I enjoy getting up in the morning. It is a big part of the legacy I would like to leave behind when I am gone.

Your mission might include being the best wife and mother—or husband and father—you can be. Or it might be to build positive win/win relationships both at work and home. Part of your mission might be to become proactive in all things in every possible way. Or it could be to become proficient at something new every year or to learn something new every day.

Your mission statement should become a roadmap of what to do—a *How To* of how you could and should go about achieving your vision. Anyone who reads it should at least be able to ascertain what *you* think your purpose in being here is all about! It should be an outline of what you should be doing—need to be doing—in order to become a better you. Is your plan to create a better reality for both today and tomorrow? In other words, it could be just about anything that moves your life in a positive direction and brings you somehow closer to achieving your vision of the future.

Once you have established a personal vision of the future and developed a mission statement designed to help you focus on making that vision come true, you can begin working on personal goals and objectives. The key here is to always be sure that your personal goals and objectives are aligned with your personal vision and mission, and that your vision and mission are consistent with your values and guiding principles. The time to start thinking about vision and mission statements for your business is *after* you have put your personal house in order—not before. Otherwise, you run the risk of conflicting goals and objectives.

Defining Your Vision

The exercise you find yourself involved in when creating a corporate (or business) vision statement is similar in many ways to what is involved in determining your personal vision. It involves visualizing your business as it could be—and probably will have to be—in the future, just as a personal statement of vision or mission places you in the future with your wants, needs, and expectations realized. The difference is ***consensus***.

> **con·sen·sus** (kən-sen'səs) *noun* **1.** An opinion or position reached by a group as a whole or by majority will: The voters' consensus was that the measure should be adopted. **2.** General agreement or accord: government by consensus.

When you approach a personal vision or mission statement you have to be aware of your relationships and responsibilities, but the vision you have for the future is essentially yours and yours alone. You may place yourself in a satisfying and fulfilling future with regard to those close to you, but they don't have to come along if they choose not to. In the end, the vision is really yours—not theirs.

In business, a vision must include everyone actively involved in your business—*everyone*. Exclude anyone, and that one person's inability or unwillingness to share your vision can and often will subvert all your hard work and effort.

In a section titled "A Vision Provides Purpose And A Sense Of Mission," in *Counterintuitive Marketing,* the authors describe corporate vision this way:

> *A vision, at its most basic, dictionary level, is, amoung other things, an ability to see something not visible. It is a force or power of imagination, something supernaturally revealed, as to a prophet. A vision provides a company with a purpose and a sense of mission. Visions define a few outstanding goals around which companies can organize their resources; they help to inspire the workforce to pursue common aims.*
>
> *Most companies state their objectives, but a challenging objective, though a key component of a vision, is not the same as a vision. A vision is a hope, a goal, a dream; it incorporates the values of the company (or entrepreneur), and it implies benefit for customers. A powerful vision looks outward — or the company is in trouble. A vision statement expresses the end. It expresses what the company wants to be in the future.*
>
> *The vision must be exciting, even inspirational, to all the company's stakeholders: investors, the general financial community, employees, suppliers, prospective customers–everyone. The vision must be so big, so bold, and so ambitious that expressing it–never mind executing it–has a transformational effect. The company starts to become what it wants to be in the future.*

I have read scores of books that spend hundreds of pages trying to express the same elegantly simple idea just as articulately without doing it as well.

Your corporate vision should be capable of inspiring those around you to perform at the highest level of their abilities—it should excite you and everyone around you. It should be clearly articulated and very specific. Anyone who reads it should be able to understand where you are going and why you think the journey is worthwhile. It should also be unique, special, and different. It should suggest leadership in your market or state clearly that leadership and domination is where you are headed. And, it should make anyone and everyone that reads it want to associate or align with you, your company, and everyone associated with it.

It asks and answers a number of key questions, like: Who are we? What do we stand for? Who are our customers and how can we serve them better than anyone else can or will? What are our guiding principles? What is important to us? And, finally, can we really do it? Can we succeed? Can we make it happen?

The vision statement we developed looks something like this:

> Schneider's Automotive Repair, Inc., will be the premier supplier of high-quality automotive products and services.
>
> Our future will transform our customer's reason for doing business with us from purchasing automotive parts and services to purchasing confidence and security.
>
> We will be recognized in our area and throughout our industry as knowing our customer's wants, needs, and expectations as well or better than they do.
>
> We will be a partner to our customers and provide value-added products and service in anticipation of their needs.

> We will continually improve the automotive service experience of our clients, the profits of our corporation, the growth and welfare of our people, the image of our industry, and the betterment of our community.

Vision is to leadership what mission is to management—the difference between doing things right and doing the right things. Vision is where you want to be, while mission expresses exactly how to get there. In fact, a clearly stated vision statement will almost define its own mission statement. It is the official roadmap for your journey, and the clearer the vision, the more accurate the map.

There is one other thing to consider before we move on. It should be assumed that the company's vision will be consistent for an extended period of time. Your mission may change year-to-year, or even quarter-to-quarter, depending on any one of a number of external stimuli. Vision, however, should not change capriciously—in fact, it should remain relatively constant unless there is some profound event that forces a change, such as a dramatic change in technology, a profound redistribution within the market, or perhaps arriving at your vision. It should take you far enough into the future to make the events of doing business day-to-day almost irrelevant—out of focus, almost invisible.

UNDERSTANDING YOUR MISSION

What is your vision for the future? What is your mission? What goals do you hope to achieve? What are your objectives? What principles would you like to guide you and everyone else in your organization?

In developing a powerful mission statement capable of inspiring great performance, you will have to both ask and answer a number of key questions as well—questions like: What value-added product or service do we bring to our industry and our customers? Do we understand the problems our customers bring to us every day? Do we have the solutions they are looking for? Do we understand how our company is perceived in the market place? Do we understand our position in the minds of our clients and communities? How are we actually going to achieve this vision? What internal and external resources will we be able to bring to bear? How do we meet or exceed our customer's wants, needs, and expectations? Can we do it consistently? and How can we do it better?

Our mission statement developed over time and appeared on a number of our promotional pieces and advertisements before we ever realized it was a mission statement at all. It stated what we wanted to accomplish clearly enough. We still never thought of it as a mission statement until we realized just how hard it would be to come up with something that reflected what we were trying to accomplish any more clearly—any more passionately. Simply stated, our mission reads as follows…

> To reinvent Schneider's Automotive Repair into a Total Quality Service-oriented, customer-driven company focused on providing quality automotive service to our customers, improved quality of life for our employees, and increased profits to our stakeholders by doing what we do best…
>
> Providing the Needed Work, Done Well, Finished on Time, and at a Fair Price, With All Work Guaranteed —
>
> Ultimately, becoming as important to our clients as they are to us!

A mission statement should have anywhere from one to five central themes along with the name of the company and a statement of *for whom* the organization will carry out this mission. In other words, a statement of who will benefit if the mission is successful.

The mission statement should include *action* words like *provide, manage, maintain,* etc. It should capture the essence of why the organization exists... and that statement of organizational purpose is—or should be—your mission statement.

Discussions of mission and vision can be confusing at times. I've seen vision statements that were really mission statements. It seems more common for the confusion to move in that direction, perhaps because it is easier to articulate what you want to do than it is to talk about what you would like to become in the future. One of the tools you can use to identify whether or not something is a part of your vision or might be more appropriately incorporated into your mission is this key question: Is what you are talking about something you are accountable for (accountability almost *implies* values) or is it a task that needs to be performed along the way? If it is task-related, it almost certainly has more to do with mission than with vision.

VALUES AND GUIDING PRINCIPLES

Before we begin focusing on goals, objectives, and tactics, now might be a good time to revisit the concept of guiding principles and the role these principles and values they represent play in defining what your company is all about. If reality is the world as it is, your values should represent the world as it *could be*, especially if those values are aligned with your guiding principles. These values, when institutionalized by a company, become the principles that guide it, and these *guiding principles* will determine just what kind of a company it will be—how it will treat its customers, its employees, its vendors and suppliers, and how it will interact with the community it serves.

The following guiding principles belong to Charles Schwab & Co. I found them in a book by Jason Jennings & Laurence Haughton entitled *It's Not The BIG That Eat The SMALL, It's The FAST That Eat The SLOW!* and they look something like this:

1. Is it fair to our customers?
2. Does it respect our fellow employees and the spirit of teamwork?
3. Are we striving relentlessly to improve what we do & how we do it?
4. Will it earn—and will it be worthy of—our customers' trust?
5. Will it reinvent the business?
6. Are we willing to risk short-term revenue to do the right thing for the customer and ensure long-term success?
7. Will we own the technology?
8. Does it leverage the brand to build trust?
9. Will it create and nurture a spirit of innovation?

These are the guiding principles of Lend Lease Corporation, found in the same text:

1. Dare to be different in everything we do—become the enemy of mediocrity!
2. Never do anything that would diminish the pride our parents have in us!
3. No nasty surprises—grow earnings every year!
4. Be a leading employer.
5. Enhance the environment.
6. We need special relationships to enhance our capabilities.
7. No individual has a monopoly on good ideas.
8. We will only prosper with the support of the communities with which we interact.
9. We would like all employees to be shareholders.
10. We believe there is a strong link between good governance and performance.

When these guiding principles are applied properly, they guide every decision made by anyone and everyone in the organization.

Where do they come from? They *should* come from you! They should flow out of the way you see your world—how you would like to be treated and how you insist on treating others.

They are a statement of how you see the world around you... good & bad, right & wrong, what is moral and what is not. It defines the role of your company in the community it serves and it recognizes the importance of all those around you: customers, employees, vendors and suppliers.

How are they enforced? Before they can be enforced, they must be observed and experienced. Leaders demonstrate by example, they don't demand. When that happens, everyone in the organization internalizes the guiding principles of the leader—the president, chief executive officer, or the individual responsible for driving the company forward—and then watch as these principles become an integral part of the company's culture. They aren't just spoken—they are lived out every day.

It begins by doing as you say—by living your principles and values—not just by talking about them. It begins with consistency, and generally ends with respect!

It begins with you. It's your company... so, it's *YOUR* job!

1. Start the process by creating and then publishing a set of guiding principles and then live them!
2. Try to make sure that all decisions are consistent with those guiding principles.
3. Make sure every one knows and understands what they are and the consequences of acting against them!
4. Make sure everyone understands they must act in accordance with them at all times.
5. Discuss them frequently.
6. Never publicly reward or acknowledge anyone who acts against them, even if their actions result in short-term gain.

Values and guiding principles relate back to your system of *enduring beliefs*... beliefs that have stood the test of time and endured throughout history such as honesty, integrity and character. They are the foundation of our civilization and act as the guidelines for mutual respect and open communication, ultimately becoming representative of the system or style of management a company will adopt.

Whether you are aware of it or not, your company operates within the boundaries of an established set of values and guiding principles. It would be in your best interest to know what they are and understand how they impact you, your company, and everyone around you.

ESTABLISHING PERSONAL AND PROFESSIONAL GOALS

If you don't know where you are going, any road will do!

And, for most people, that seems to be good enough. We are living in a culture that seems bent on burying us in clutter and distraction. There are diversions everywhere and more choices than anyone should ever be confronted with—some of them both attractive and compelling. Without some way to cut through the clutter and the noise, finding your way is almost impossible. It's like being lost in a deep fog.

A clear vision of the future, your future, and the future of your business eliminates the confusion. It gives you a destination and your subconscious mind the opportunity to chart a course there. Simply verbalizing that vision—stating it publicly—can be the magical beginning of an unbelievable adventure.

Where do you want to be in a year, two, five, or fifteen years? What will your business look like? What will it feel like when you finally get there and realize what you have built? What will it be like to work in such a place? And, perhaps most importantly, what will it be like to do business with your company from the perspective of a customer or a supplier?

If you can close your eyes and see that business as it *could* be and not as it is you are way ahead of your competition. But first you have to think about your business in ways you never have thought about it before—in the future, not in the present.

One of my favorite passages in literature addresses the issue of not knowing where you are going at its very core. Unfortunately, it is all too appropriate for the majority of people we come in contact with on a daily basis. It comes from Lewis Carroll's *Through the Looking Glass*. Alice encounters the Cheshire Cat just as she realizes how lost she has become, and asks for the cat's help. Their exchange goes something like this...

> *The Cat only grinned when it saw Alice. It looked good-natured, she thought: still it had VERY long claws and a great many teeth, so she felt that it ought to be treated with respect.*
>
> *'Cheshire Puss,' she began, rather timidly, as she did not at all know whether it would like the name: however, it only grinned a little wider. 'Come, it's pleased so far,' thought Alice, and she went on. 'Would you tell me, please, which way I ought to go from here?'*
>
> *'That depends a good deal on where you want to get to,' said the Cat.*
>
> *'I don't much care where—' said Alice.*

'Then it doesn't matter which way you go,' said the Cat.

'—so long as I get SOMEWHERE,' Alice added as an explanation.

'Oh, you're sure to do that,' said the Cat, 'if you only walk long enough.'

Unless *SOMEWHERE* (regardless of where that somewhere turns out to be) is where *you* want to be, establishing a clear set of goals and objectives to keep you on track while you are traveling will be critical to your success. Doing that effectively and intelligently is what this portion of the book is all about.

Before we do that, though, I'd like to ask you to try something with me. I'd like you to take a minute and think about your vision of the future and what that future will both look and feel like. Once you've done that, I'd like you to take another moment to think about the steps you would have to take to get that image from vision to reality. Certainly, there are specific things—specific components of that vision—that would have to be addressed. What are they and how would you go about achieving them?

Now, do yourself a favor and write them down. First, write down your personal goals (we'll get to your professional goals in a minute). If you could change anything about your life in a positive way, what would it be? What's important to you?

Would you like to spend more time with your family or would you like more time for yourself? Are you looking for financial security—something to help meet your wants, needs, and expectations for the future? Have you even thought about the future? Do you want or need more money now for your present needs? Do you dream about a new boat, another car, or maybe even flying lessons and an airplane? What about personal freedom, traveling abroad, or planning a really spectacular vacation? And, finally, what about personal development—exposing yourself to new ideas, learning new skills, and seeking out new experiences?

The answers to any or all of these questions will lead you through a process of discovery that will help you identify exactly what you need to do to achieve each specific goal. So take a minute and write down a few personal goals. Write future and present goals, as well as material and/or spiritual goals, as we discover the technology involved in changing our present—and as a result—our future.

Your Personal Goals

The process used to define your professional goals is not much different from the exercise you just went through to identify your personal goals. The same technology applies. The only difference in the process is that your personal goals are just that—personal. They generally apply to your wants, needs, and expectations, and to yours alone. The only way they impact the lives of those around you is in regard to how they impact you.

Professional goals are somewhat different, especially when we are talking about specific goals developed in pursuit of your corporate or business vision and mission. These goals require the active and voluntary participation of others in order to be realized. No matter how powerful or compelling they are to you, your odds of success are practically nonexistent if they aren't shared with all those directly or indirectly involved in the everyday operation of your business.

People will follow you, but as a general rule they will only follow you if they have some idea of where you are going. Don't forget that just because these goals are meaningful to you, they might not mean much to someone else unless you are able to communicate how achieving these goals will benefit them—how it will help them satisfy their wants, needs, and expectations, and how it will benefit them in very real and positive ways!

Aside from that, it just makes good sense. *Two heads are better than one!* And, unless you are a one-man operation, you will find yourself dependent upon the actions of other people in order to insure your success. Having them share your vision of the future—having them understand how making that vision a reality will change their lives in a positive way—is the only way you may have of insuring their participation.

Communicate your vision of where the business needs to go first. Explain how and why you feel the way you do. Tell them how important their participation is and let them (the other people with whom you will need to make the journey) know how much you want them with you when your collective vision of the future becomes a reality. Involve them in the process of expanding your vision to include everyone willing to make the journey—everyone willing to travel with you Then, invite them to help you modify your vision to include their thoughts and feelings regarding where the business needs to be and what it needs to look like in the future.

Once you have expanded that vision, start focusing on the goals and objectives required to get to where you want to be. What specific actions must be taken? What steps have to be followed?

Will you need to increase sales? If so, by how much and how will you do it? Are you looking for more income for yourself and everyone else in the company? Would you settle for the same income achieved in a shorter work week? Do you want to work fewer hours? Do you want to be there at all, or would you rather build a business capable of running without your even being on site?

I don't know about you, but two of my recurring goals are *more time to plan* and *less personal stress.* You might decide to go the other way—looking to either *grow* the location you find yourself in presently or expand to multiple locations.

Regardless of what those goals might be, take a minute to write them down.

Your Professional Goals

Goals Are Long-Range Plans

There is a hierarchy of goals that really should be explored as we continue in the course of this discussion. There are *strategic* goals, *tactical* goals and *operational* goals. Each plays a role in its own unique way. Each one helps clarify organizational imperatives.

Strategic Goals

Whether you want to position your repair facility as the premier outlet for quality automotive service and repair or as the only place to go for cutting-edge technology, strategic goals focus on the long-term direction of the company. Without clearly formed strategic goals, organizational clarity is impossible and success will be elusive, if not unattainable. At the very least, a lack of strategic goals will present countless opportunities for your competitors to leap frog over you, while you find yourself paralyzed over which decisions make sense for you and your company.

Tactical Goals

Tactical goals are goals that center on the *how to* of achieving the strategic elements of your vision. If our strategic goal was to *achieve a position as the premier outlet for quality automotive service and repair or as the only place to go for cutting edge technology,* our tactical goals would direct us to find the highest-quality automotive parts and accessories available anywhere, in addition to recruiting and retaining the best-trained, most competent technicians and then furnishing them with state of the art equipment.

Operational Goals

Your operational goals will be targeted on whatever it takes to facilitate your tactical goals. Using the two examples above, you could not make an intelligent decision with regard to which suppliers to choose without comprehensive and relevant information about these different companies, the product lines they carry, *their* vision, mission, goals and objectives, and *their* guiding principles—the way *they* do business or the way *their* customers perceive doing business with them.

Organizational Clarity

No organization, regardless of its size, can sustain high-performance success without achieving at least some degree of organizational clarity (a clear sense of direction augmented by discipline, determination, and a clear awareness of purpose.) The correlation between the absence of vision, mission, goals, objectives, and failure—or at least incomplete or unrealized success—is too high to discount or ignore.

Neither can we ignore the fact that each goal and objective we establish by its definition, the problem it solves, or the direction it takes us will be implicitly or explicitly tied to one of the previously mentioned categories goals (strategic, tactical, or operational).

If we are unable to recognize to which of the three categories that particular goal or objective is oriented, we run the risk of improperly forming the goal itself and subsequently subverting our own success.

This is an important point because there are few organizations that have such a singularity of purpose. For most companies, moving closer to your vision will require a number of goals addressing a number of different issues, and that will require sharing available *corporate* resources. In many cases, having more than one goal and limited resources will force you, or anyone else involved in the process of goal formulation, to prioritize and rationalize your choices based on how close or how fast they move you toward your vision.

It is not uncommon for goals—even carefully thought out and well-defined goals—to be somewhat vague. We live in a constantly changing world and that almost presupposes some degree of ambiguity. Consequently, it becomes the specific objectives that are developed to reach those goals that will define precisely what gets done, by whom, why, when, where, and how.

To move closer toward organizational clarity and effective goal planning and management, try asking and then answering the following questions:

1. How does this organization define excellence?
2. Have we discovered and do we understand our corporate values?
3. Do we know what we have to do to achieve excellence in everything we do?
4. Can we translate knowing what we have to do into specific action steps, and their appropriate deadlines in the form of goals and objectives that will move this company in that direction?
5. Can we develop a set of objectives for every goal?
6. How will we know what kind of progress we are making as we try to achieve each goal?
7. Have we developed a way to measure our progress? Do we have the metrics necessary?
8. Will monitoring our KPIs help?
9. Are we monitoring the extent to which achieving these goals and objectives are consuming available resources?
10. Is the benefit worth the sacrifice?
11. Is there a way of knowing which goals and objectives are synergistic and which create conflict?
12. Do we have a champion to monitor, manage, and mentor this process?

The importance of this last question cannot be overemphasized. Without a champion, the goals are likely to be forgotten or at least superceded by something *more pressing*. The fact of the matter is there is probably very little that is actually *more pressing*. There has to be someone to quietly—and sometimes not so quietly—remind everyone involved of the important work that has been started… someone to cheer everyone on, and recognize those involved in making the process a success.

The reality of life is such that each of us should embrace this exercise. The reality of life is such that each of us could probably benefit greatly from doing just that. However, the reality of life is also such that few if any of us will actually do that because we don't know what to do or where to start. The balance of this text will be dealing with exactly that while we try to identify the technology necessary to discover, clarify, and achieve the right goals and objectives.

GOAL SETTING TECHNOLOGIES

Like most converts, I never believed any of the *stuff* I read about the value of setting goals—written or unwritten. In fact, there was a time I might have even volunteered to lead the *Whaddaya Mean, Goals and Objectives? You Don't Need to Write Anything Down to be Successful!* parade. But, that was then and now is now! You can call it an epiphany, or you can call it responding to reason and the weight of empirical evidence.

The bottom line is simple: I've seen just how powerful this stuff can be and now I'm a disciple! It works… but you do have to realize and understand that within any technology there are laws and/or principles that govern the *way* it works.

Here are a few of the laws and principles relating to goals and objectives that I've encountered since my conversion.

Principles of Goal Setting

Written or unwritten, goals are absolutely necessary… however, they are equally as limiting.
What do I mean by that? If there are few among us with the discipline to set goals and the determination to reach them, there are fewer still who would continue on to accomplish more once that goal was achieved. Human nature would almost certainly dictate stopping for at least for a little while to reflect on the accomplishment. Aside from that, man has always been limited by his own imagination. The technology of setting goals is no exception to that rule. We will never know what we are truly capable of—what we could accomplish—because the majority of us suffer from anemic or atrophied vision.

Goals must be reevaluated constantly… They are never the ends themselves. They are only the means to those ends.

Goals bring with them the possibility they will not be completely achieved, and that's OK
Not long ago, I remember setting what I thought was an almost impossible goal for our business. I wanted to double sales within a five-year period or increase sales by almost 20 percent per year for each of those five years. That would have meant going from approximately $400,000 in sales to over $800,000 in sales. I'm almost embarrassed to say we didn't make it within the period of time I had allotted, but that's the truth. We didn't. For whatever reason, we fell short of our goal by almost $60,000… but, we got $320,000 closer than we would have gotten if we didn't have something to shoot at!

Goals must be realistic.

You have to believe your goal is attainable, which means that it is possible and that you can achieve it. That doesn't, however, mean that it has to be conservative. It can be absolutely outrageous! You just have to believe with a perfect faith that no matter how outrageous it might be, you still have the power to make it happen!

Goals must be explicit.

That means they have to be written in great detail. The more clearly defined they are, the easier it will be to achieve them. The subconscious mind is an incredible piece of work. With everything the subconscious is capable of, the one thing it is not capable of is differentiating between a perfectly conceived illusion and reality. The more detail you add to the imagery, the more likely the subconscious is to determine that whatever it is you are wishing for is something you should already have. Then it will turn all of its energy and resources—all of *your* energy and resources—toward accomplishing whatever it is it thinks you need to do to get whatever it thinks you need to have.

Goals must be written.

I know… I know… I may have implied that goals could somehow go unwritten, and I'm sorry. There was a time when I felt that way, but that is no longer the case. Not only do I believe they must be written, but I believe the less personal of your personal goals should also be posted publicly somewhere so that you and those close to you cannot help but be aware of what you are trying to do. I say this for two reasons: first, so they are reinforced by constant confrontation, and second, for the pressure that will result from publicly posting them. There is no better motivation than contemplating the disappointment those close to you might experience if you are unsuccessful!

They must have a deadline.

They say that a goal without a deadline isn't a goal at all. It's a dream. And, as we all know, dreams have no substance. To make that goal real, determine how much time you will dedicate to making it real and then get to work!

Goals must be reinforced daily.

You should have your goals written down and kept close for your own review, and you should review them at least once a day.

I believe very strongly that any action you might take that does not move you somehow closer to your mission, vision, goals, and objectives must be considered suspect. And that means that if you and I are doing something that is not moving us closer to our dreams, we really ought to be thinking a great deal about why we're wasting our time doing it.

Goals must remain fluid.

The world around you is constantly changing. Consequently, your goals and the objectives that support those goals must be flexible enough to change with that changing landscape. You can't be married to a set of goals that are no longer capable of taking you where you need to be.

You must have more than one goal.

There is virtually no such thing as a vision so singular in purpose it need only be supported by one goal. Life is not that simple. Man is not that simple. The world that confronts all of us is not that simple. Consequently, having only one goal is not that simple! You function in more than one area of life—create at least as many goals and objectives as you have roles.

An Action Vision

Visualize your life in the future, as it would be if you had already achieved all of your goals. The *action vision* of that future should be very consistent with the statement of vision you created to get there. Can you see and feel yourself functioning in that future? How clearly can you see it?

What kind of business will your business be? What kind of cars and trucks will you be working on and what will your customer base look like?

Create an action plan or a *strategic plan* whose goals and objectives will move you one percent closer to your goals every day.

Where Would You Like To Go Today?

One of the reasons more people aren't developing a dramatic vision, creating compelling missions, or identifying meaningful goals is the fact that we make it look and feel so much more complicated than it is. I know that's how I felt. The idea of figuring out where you want to be in the future when every day in the present leaves you feeling as if you aren't really sure where you are, isn't always very appealing! Then, there is always the pressure you may bring upon yourself if or when you fail or fall short.

Regardless of how hard or easy we may think it looks, setting goals is really not all that difficult. The bottom line is: it's a lot simpler and easier than you may think.

Where Are You?

Like so many other things, it starts with a question… and, the question is: Where are you? We've been discussing the importance of vision, mission, goals, and objectives, but these powerful tools are only important if we harness the power they afford us and allow that power to propel us into a better, more satisfying and fulfilling future. In order to do that, you have to have at least some idea of where you are before you start thinking about where you would like to be.

Establishing a baseline will provide you with the measurement metrics you will need to both monitor and judge how you are doing in your efforts—how far you've actually come with regard to where you started.

It does something else for you—something that goes unmentioned far too often. It forces you to take a minute to confront where you are at that particular moment in time and possibly realize just how far you may have already come.

Where Would You Like to Be?

The last few chapters have been all about determining where you would like to be. That's just about all we've been talking about. The second, and perhaps most important, step in planning the journey is picking the destination. What is your vision for the future? What is your vision for your company and the people close to you in that future?

Remember, to be successful this has to be—*must be*—a clear and vivid snapshot of you and your company, fully functional in a fulfilling and satisfying future.

How Are You Going to Get There?

In a few pages, we will begin discussing the steps required to properly plan such a journey. Until then, let's just say that it behooves anyone contemplating a trip of any kind—let alone an odyssey like this—to plan extensively for any possible contingency. The plan must be detailed and explicit, complete with each person's individual responsibilities clearly spelled out.

How Will You Know When You Get There?

Having base-lined performance by freezing your business in time with the KPIs we discussed in Chapter 3, we can safely establish a picture in both words and numbers of what *where we want to be* looks like. We'll know we're in the right place because we can measure it. And because we can easily measure the distance between where we are and where we want to be, we can establish our arrival (just as easily as we can follow driving directions that begin with *drive east 2.7 miles* after we have reset the trip indicator in the family car and checked the compass).

More Questions... Then Answers

Everything I've read and most of what I have experienced personally tells me that being successful in business is really all about curiosity as much as it is anything else. Success requires an inquiring mind, or at least a sense that as good as things might be, they're not as good as they could be. That leads you to wonder Why? and What? ...and then ask How? ...things like what happened, why did it happen, and what we can do to do more of what works and less of what doesn't.

It is a healthy exercise. In fact, questions like *How*, *Why*, and *What* fueled the quality revolution.

The same dynamics hold true for your values, guiding principles, vision, mission, goals, and objectives as well. It certainly doesn't hurt to question your beliefs from time to time, especially in a world that is constantly changing all around you. At the very least, it will leave you sure that what you believe is *right*—if for no one else, then for you. Consequently, I would recommend that you ask the following questions once you have established a set of goals you are about to embrace for your business.

Why Do You Want to Achieve This Specific Goal or Objective?

There really should be a compelling reason to take the time and make the effort to achieve whatever goal you are pursuing. It might be a good idea to isolate that reason and understand its implications. Will achieving this goal really move you closer to your mission or your vision? How? Why?

Clarifying the answer illuminates the question. You may just find out that achieving that particular goal brings you no closer to your strategic objectives than you were before you started.

What Are the Specific Benefits This Goal or Objective Will Produce?

It's always good to know what will come out of the sacrifice and effort required to pursue any goal or objective—what the ratio of investment to reward will be. It doesn't hurt to explore the answer to that question before you dedicate the resources required for success, either. If it's a business-related goal or objective, the targeted results should be very clear. If they aren't, you have to ask yourself how you will measure success.

Who Will It Benefit and How?

There should be some positive benefit that results from the accomplishment of a goal, or even an objective. Identify the benefit and then follow it back to the beneficiary. If it is a corporate goal, everyone should benefit in some way—owners, managers, technicians, support personnel, customers—*everyone*. Along with the who, you should also take a minute to look at the *how*.

Are These Benefits Important to the Important People Around You, or Will They Benefit Just You?

Business goals that only benefit the owner or the managers are doomed to failure. Being successful in business means soliciting and obtaining the active and voluntary support of all those involved in your business. If those individuals cannot identify the WIIFM (the *what's-in-it-for-me*), their support is likely to be anything but *voluntary* and *active*!

What Price Will You Have to Pay to Achieve This Goal or Objective?

Don't lie to yourself or anybody else. There will be a price—a price in time, dollars, or in a sacrifice of one kind or another. Anything you choose to do will be done at the expense of something else. In other industries, it's called *loss of opportunity*. I do presentation work all across the country. Accepting a speaking engagement from a sponsor means that date is no longer available. It's off the calendar. No matter what comes up after that date is scratched off, regardless of how attractive or financially rewarding it might be, the opportunity is lost because of a prior commitment. So look carefully at what you will be giving up to accomplish each of the goals you have chosen to pursue. In some cases, the price you pay will far outweigh the potential benefit.

What Are the Benefits of Not Setting This Particular Goal or Objective?

Be honest! There are benefits to not setting a goal or an objective. You won't have to deal with the discomfort of confronting the unknown. You won't have to deal with the uncertainty of change. And you won't have to deal with the rejection either (the rejection of those close to you feeling threatened by your new behaviors and the new direction of the company).

What Is the Cost of Not Setting This Particular Goal or Objective?
What is the cost of not being proactive…

In a world that is moving as quickly as ours is, standing still is the same as falling behind! The cost of not setting goals or objectives for your business is familiarity, complacency, and a slow, painful death by attrition. Even the wrong direction is sometimes better than no direction at all. However, it is advisable to spend a little time thinking about the results of not setting the wrong goal or objective, or a goal or objective that is not clearly defined, and what those results would cost in time, money, lost opportunity, and misappropriated resources, etc.

ESTABLISHING PERSONAL AND PROFESSIONAL OBJECTIVES

What do you have to do to transform your vision into reality? You develop your vision to provide meaning, focus, and direction for the everyday activities of running a business in a changing world. In order to be effective, this vision must reflect the organization's values, guiding principles, current reality (a business has a current reality just as an individual does), and environment. If the world in which we operate was static and unchanging, there might be no need for strategic planning, goal setting, or a vision of the future. But that isn't the case. Virtually everything about our world is in a constant state of change, ranging from the demographic, economic, and psycho-graphic behaviors of our customers to the social and cultural values of our society.

Later on, we'll be taking an in-depth look at how all of this comes together in a strategic plan. But, for now, I'd like to focus on the process of establishing timely, realistic, and achievable goals and objectives as an integral part of transforming your vision into your current reality.

You achieve your vision through the relentless pursuit of your goals and objectives. To do that, you must identify your goals and establish the appropriate objectives. It may sound easier than it is, but it can be easier than most people make it out to be!

Once you (a *collective you*—including *all* individuals critical to your business success) have established a vision, formulated a mission, and created a strategic plan reflecting both the company's internal and external environments and focusing on the changes necessary to meet current and future needs, it's time to move your attention to the incremental and concrete steps necessary to make it all happen. This is a fascinating exercise that will demand a realistic evaluation of your capacity to sustain the changes you are about to invoke. At the very least, it is a process of inquiry and evaluation—discovery and assessment.

Five W's and an H

In a discussion of strategic planning, the five W's represent the *who, what, where, when* and *why* of setting goals and objectives. The H relates back to *how* you are going to get things done.

Because goals and objectives are established as an integral part of a strategic plan to accomplish a specific mission, the most critical of the five W's to focus on should be *why*—Why are we doing this?

Your *why* should be the guiding force in reconciling wants with needs—the allocation of resources with functional or organizational demands.

The objectives you develop are the action steps required to accomplish those goals. They refer specifically to *who* will be responsible, *what* exactly needs to be done, *where* you will have to focus both your efforts and your resources in order to accomplish this goal or these objectives, and *how* and *when* all this will need to come together (what is the timeframe involved).

The acronym I've seen most often associated with any discussion of goals and objectives is SMART. To be effective, goals and objectives must be SMART—Specific, Measurable, Achievable, Realistic, and Timely.

> **S**PECIFIC — individual steps required to achieve a particular goal or objective
>
> **M**EASUREABLE — metrics involved in determining how close or far away you are
>
> **A**CHIEVABLE — possible realizing the present capacity and resources of the company
>
> **R**EALISTIC — possible given the current resources of the company
>
> **T**IMELY — accompanied by recognized deadline

These goals and objectives will form the skeleton of an *action plan* (a plan that specifically addresses how each strategic goal will be met): which objectives are appropriate, how they will be applied, by whom, the appropriate timeline, and how each goal and/or objective will benefit the enterprise. It should also include the metrics that will be used to measure and monitor both your progress and your success.

Another acronym that might help you to associate what we are discussing here with your situation is SSORT (pronounced *sort*) ...as in, how will you SSORT things out?

> **S**TRATEGIC GOAL — your goal
>
> **S**TRATEGY — which goal or objective will be addressed first
>
> **O**BJECTIVE — first objective to reach in order to start the process
>
> **R**ESPONSIBILITY — who will be responsible for accomplishing the objective
>
> **T**IMELINE — what is the established time frame

Process

Someone once said that *if you fail to plan, you plan to fail.* I'm not sure anyone could effectively argue with that. Without planning, your dreams will remain exactly what they are—dreams. Planning, accompanied by appropriate action, is what moving from dreams to reality is all about. They substitute precise, detailed, thoughtful, explicit action for a vague and fuzzy aimlessness and lack of direction. They also provide the appropriate benchmarks for measuring orga-

bench·mark (bench'märk) *noun* 1. A standard by which something can be measured or judged

nizational performance. Furthermore, companies, managers, and even individuals who have established and then aggressively pursued specific goals and objectives have typically outperformed individuals functioning without such a plan.

As we will see, goals and objectives established for your business can vary between long- and short-term, financial, and operational. They can be general or strategic and very specific. Without the appropriate focus, we are back on the road with Alice realizing that the Cheshire Cat was right: *If you don't know where you are going, any road will have to do!* With the proper vision, a clear mission, and the appropriate goals and objectives, we can prioritize between the pressures of short-term financial reward and long-term strategic growth and market dominance.

Examples of financial objectives might include achieving growth in revenues or profits within a certain percentage or within a fixed period of time. Or, it could be increasing gross or net profit margins. Examples of what your strategic goals or objectives might look like include greater share of market or greater share of customer, broader product offerings, or the development of additional profit centers (i.e., tire sales for a facility not presently involved in the sale of tires), reinforcing the company's position in the minds of its customers, increased customer service, or the establishment of the company as a leader in a certain service category.

What we are really talking about here is *strategic intent*—the level of commitment to achieve established objectives beyond the company's current reach and resources.

This strategic intent is what will be necessary to drive the company forward in the long term (three to five years), and the short term (within one year). Short-range objectives clearly define the immediate results necessary for success as well as the timeline and level of performance required. They are generally sequential in nature with one objective following another. Long-range objectives require high priority actions in the present to help reach more distant—but no less important—performance goals and objectives later on.

There will always be a difference between the level of performance generally perceived as *possible* and the level of performance possible when the organization is challenged and all available resources are committed. The real challenge is to consistently achieve that level of performance all the time!

Analysis

The need to establish goals and objectives is clear. The technology involved isn't clear. I suggest you start with a series of simple questions to help you analyze your current situation, as well as your future needs.

Where are you going with this company? Is one of your goals total quality automotive service and repair? How will you accomplish that goal? Will it take more technical training? If it will, where will you get that training? High school and community college vocational education training programs are evaporating, and manufacturers are increasingly less able to provide such services.

Is there a problem with the parts you are currently securing for your customer's cars and trucks?

Is the quality suspect or is their service or their ability to fill your orders 100 percent effectively a problem? Do you need to source different lines from new suppliers? If you do, who will you look at? How will you find them?

Do you need more help? Perhaps you need different help with different skill sets. Will you need to reevaluate your compensation and benefit plans to keep the technicians you have and attract the technicians you need?

Will education and training focused on delivering better customer service move you closer to your goals? If it will, who will receive this training? How and where will you access it? Perhaps, as important, how much will it cost? Where will the money come from to pay for it all?

Your problems could all relate back to equipment or facilities management challenges... Maybe the time has come to re-capitalize with state-of-the-art equipment. Or maybe we have to reevaluate how we deliver our products and services?

Are satisfied customers our goal? Do the products and services we make available to our customers meet or exceed our customer's wants, needs, and expectations? Do we know what those wants, needs, and expectations are? If we don't, how can we find out? If we fail to do that, there is no way we can hope to satisfy our customers.

Each question demands an answer and each answer is the birth of a strategic objective poised to move us one step closer to our goals—one step closer to the realization of our dreams.

Application

Let me share a personal example of how all this fits together before we move on. A number of years ago, we found ourselves struggling with what we felt was an untenable *current reality*.

My business is a family business, and my parents and I have been eating breakfast together for more than 35 years. A number of years ago, while waiting for my parents to join me and reflecting on the business, I scribbled a note on a napkin. At the time, business was good but relatively flat. Profits were adequate, but not growing. Our marketing efforts were satisfactory, but not much more than that. This is what I wrote:

> Last year's advertising budget: 0.5% of Sales.... Not enough. 10% increase in productivity = $10,000+ per man... assessment of each person in the shop. ... look at the different kinds of work and what we were sending out... need to recapitalize investment.... remodel to bring the visual presentation of the shop in line with our image of what we want to be... equipment needs... staff... needs... policies and procedures... need for freedom to focus and concentrate on the business.... Note that PepsiCo has just spun off its fast food restaurants into a new division to concentrate on core business needs... need to focus on core competencies and doing what we do best... and outsourcing... buying what you need out.

That was the beginning of a new vision of the future and consequently, a new mission. Each one of those scribbles became the focus for a new area of concentration—a new area of concern. Subsequently, each resulted in new strategic direction for the company accompanied by new goals and objectives.

The first scribble recognized that our advertising/marketing budget was sadly inadequate. One-half of one percent of sales spent on advertising for a business doing more than $660,000 in sales is nowhere near enough to create the kind of image and awareness we were looking for. Translated into dollars, you would have to ask yourself how many envelopes, stamps, and pieces of paper would that half of a percent (or $3,300) buy? How much time for someone to fold, stuff, and stamp them? How many service reminder postcards would that equal? How many promotional pieces?

A number of business analysts have suggested that it takes two to five percent of sales to grow your business and have your message heard above the clutter and noise created by everyone else competing for a customer's attention. You can buy exponentially more awareness for $13,200 to $33,000 ... if you have a plan ... if you understand whom you are targeting and what your segment of the market looks like.

The **goal**—our strategic imperative—became increased image and awareness translated into increased traffic (higher car count by 2 cars per day) and sales (ten percent increase per year for five years).

Our first **action step** was to increase the advertising/marketing budget to 2.5 percent of sales, or $16,500. This budget would be distributed as follows (a new series of specific **objectives**): customer follow-up phone calls to be performed by existing staff (no real increase in existing cost), a quarterly marketing piece mailed to existing customers with a history of more than $1,000 in purchasing per year (budgeted at $12,000—$3,000 per quarter or $1.00 per piece), and the balance of the marketing budget reserved for service reminder postcards and miscellaneous promotions.

It doesn't sound like much, but compared to what you can buy in the way of effective marketing and advertising for $3,300, it's more than a start. When it moved off the napkin and into a marketing plan, it started to look something like this:

Goal: Increase Image & Awareness

Strategy: Get our name in front of the customer in print, through the Internet, or by telephone once every six to eight weeks.

Objective: Greater reach, increased frequency

Action Steps:

1. Increase Advertising/Marketing Budget to 2.5% of last year's sales
2. Begin calling customers within two weeks after service for purposes of quality assurance (no additional budget required)
3. Design and distribute quarterly marketing piece—budget $3,000 per quarter (keep cost per unit below $1.00—2,000 pieces to targeted members of the regular customer-base, 1,000 to selected potential clients specific to your target market).
4. Reserve balance of the budget for service reminders and miscellaneous marketing efforts consistent with our plan.

A second goal implicit in the napkin scribbles was a ten percent increase in service bay productivity. Service bay productivity is a subject covered in much greater detail in another book in this series, but we'll take a minute to go over why it was specifically addressed on the napkin and the subsequent objectives that seemed to flow out of that goal.

Goal: Increase Individual Productivity by ten percent.

Strategy: Increased productivity = increased opportunity: more work on fewer cars in the same fixed period of time.

Objective: Increase labor and parts sales through productivity.

Action Steps:

1. Reduce lost productivity by managing the flow of parts to the technician more intelligently by:

   a. supplying better, more specific vehicle information

   b. ordering earlier in the process

   c. thinking ahead to include parts possibly needed for ancillary service operations (thermostat replacement w/water pump)

   d. inspecting parts immediately upon delivery to insure they are correct.

2. Needs assessment for additional tools and training and equipment necessary for higher productivity.

3. Quality assurance to reduce recidivism (return work or comebacks): post-repair inspection and road test

4. Introduction of a repair order rack system to reduce wasted time and unnecessary communication

In the automotive service environment, productivity relates to the ability of the technician (or all the technicians in a shop) to produce billable labor hours in relation to the total number of hours available. In other words, 100 percent productivity relates to the ability of a technician to bill or invoice eight labor hours in an eight-hour workday. Productivity defined as such has two distinct components: first, is the ability of the technician to perform the work in the time allotted, and second, the ability of management to sell the appropriate number of hours available.

One is the concern of the technician, as long as he or she is provided with the facilities, tools, training, and equipment required to complete a task in the appropriate time. The other relates to the ability of the management to sell, monitor, and manage the time the technician is available for productive work.

As a whole, increased productivity brings with it higher labor sales, more productive labor in the time allotted, as well as the sale of additional parts and accessories that would normally accompany such labor sales.

The unstated, and yet accepted, average productivity for our industry has traditionally been below 55 percent. That translates to 4.4 hours billed or invoiced for every eight hours on premises. Or, perhaps stated more dramatically, 3.6 hours of wasted time every workday! A ten percent increase in productivity increases billable hours by .44 hours per day, or more than two hours per week. That may not seem like much at first glance, but let's see what happens when we extrapolate those numbers to where they logically must go.

As we just saw, a ten percent increase in the industry average of 55 percent results in .44 more billable hours per day, or more than 2.2 billable hours per week. If our labor rate is $65.00 per hour, those 2.2 hours equal $143 per week or $7,150 per year (based upon a 50-week work year, reflecting both vacations & paid holidays) in labor dollars and an additional $7,150 in the parts sales normally associated with the sale of labor or time. That $14,300 is accompanied by no additional cost, except for the additional cost of direct labor (what you actually pay the technician), and if you are paying your technicians by the hour or by the week, you don't even experience that. Consequently, the increased profits that follow increased productivity go almost directly to your bottom line!

By monitoring and managing the parts process more intelligently, and through training, facilities management, and the repair order rack system that we integrated into our operation, we saw our productivity soar from just above industry averages to just under 100 percent.

These two specific examples of where goals and objectives come from and how they can be applied should serve to underscore just how critical this process is. Without these specific goals and objectives—without a plan—this kind of success or performance would have been unlikely, if not impossible.

Beware The Danger Of Incremental Change!

I would like to interject one caveat here. It has a lot to do with vision, mission, goals, objectives, and strategic planning. It has even more to do with how and where you focus your energy and your efforts. Most change comes from the realization that things cannot remain the way they are—especially, if the pain associated with where you are is perceived to be greater than the pain associated with wherever it is you are going.

To better serve customer wants, needs, and expectations in the future, we should probably contact those customers in order to find out what wants, needs, and expectations could be better served and how. There is a danger here… something not everyone is aware of… something not everyone will warn you about.

When asked, customers can and will only respond to survey questions regarding the nature of your customer service or their satisfaction with the services your company provides, in terms of something they know—something they have already experienced. Consequently, the answers garnered from those questionnaires aren't always helpful or effective. Because the answers to those questions are related to the perception of specific services, the answers will be tied to the questions. In other words, the customers you ask about the quality of your service delivery will tell you what you want to know the only way they can, by relating your question to *the quality of service they have experienced while in your care and custody*. And, while that might seem like that's what you're looking for, I'm not quite sure it is.

If you are really serious about delivering world-class service—service like nothing else the customer has ever received before—it's not incremental improvement you're looking for, especially when everyone else in your market segment is trying to do the same thing. So, the real question shouldn't be *How do I improve the quality of my service?* or even *How can I improve the quality of how that service is delivered?* Both of those questions will lead both you and the customer being surveyed down the same path of incremental improvement!

The real questions should be:

> *What do you really want from this relationship? What do you really want to come out the other end of the automotive service tunnel and what should it look like? What should it feel like to you?*

That's where you will get the kind of answers that transform your operation into a customer-focused, customer-centered organization driven to provide world-class service and customer satisfaction.

SUMMARY

Nothing of consequence happens in our lives until we determine that where we want to be presents a better alternative than where we are. When that happens—when our vision of the future is powerful enough to overcome our fears and insecurities—our current reality will shift. For us to help manage that process, we must develop a clear vision of that future and a strong sense of our mission, as well as a way to develop the goals and objectives required to achieve both.

Take a moment and close your eyes. Think about your life and your business as you would ultimately have them both blossom and unfold.

What would that life be like in that future? What would the changes that would have to occur in order to enable such a life mean to you and to those close to you? What would your business look and feel like? What would it be like to own and manage such a business? And, what steps would you have to take to make it all come to pass?

Write your answers in the chart on the next page.

Thoughts

Actions

Results

CHAPTER 5

If You Fail To Plan

INTRODUCTION

In this chapter you will become familiar with the basics of *profit planning* while you learn how to:

- Use a personal profit demand spreadsheet
- Use a reverse budgeting spreadsheet

PLANNING BACKWARDS

How about something different for a change? How about working *backwards* to your goals and objectives?

It seems odd to offer you a buildup like the one in the previous chapter and then start talking about *planning backwards*, but that's exactly what we are going to do. You see, there is a critical question that gets left out of almost every discussion of shop performance I've ever seen, and that is… What do you want out of all this?

That's right! What are *you* looking for?

Really… if you have money to invest in creating the kind of automotive service environment you dream about, what kind of a Return on Investment (ROI) would you be looking for?

If you own a shop, take a look around you. How much money have you got tied up in inventory and equipment? I'll bet it's substantial! And what about the time you spend there—the effort and the energy? What's all that worth to you? Perhaps a better question would be: How much are you paying yourself to work in your business? Is it enough? How much would you have to pay someone else to do what you do? Would it be more? If it is, you probably ought to ask yourself why your efforts and your time are worth so much less?

If you put your money in the bank—in a savings account, a Certificate of Deposit, or in mutual funds—what kind of return would you be getting? Are you getting at least that much today? Is what you are getting today good enough for tomorrow—good enough for the future? If it's not, you've got to ask yourself another question... Why not?

If you really did ask a group of shop owners or managers for the answer to the list of questions above, the most common answer would probably be No!, Not enough!, or Because I didn't think about it! I know, because I've done just that.

The answer to these questions and a lot more like them is important. The answers are important because they are the difference between creating *a business that works for you* and *creating a place you can go to work every day*! And, believe me, they are not the same thing!

Most entrepreneurs go into business out of frustration—as the result of a blind rage or on a whim. Michael Gerber, author of *The 'E' Myth*, calls the driving force that fuels most small business start-ups *an entrepreneurial seizure*, and that's just about what it is. But it doesn't have to be that way. Just as you can plan to better serve the wants, needs, and expectations of your clients, you can create a business capable of meeting *your* wants, needs, and expectations. But you'd better have some idea of what those wants, needs, and expectations are going to be first. Once you've established that, the rest is just a matter of doing the math.

Reverse Budgeting

Reverse budgeting has been around for a while, although it seems to have eluded a great many people in our industry for a long time. The concept is deceivingly simple—you develop a budget backwards, starting with what you would like to earn first, then creating the performance goals you must set in order to make that goal a reality. In many ways, it is a direct extension of everything we've been talking about so far. You start with a *financial needs assessment* (also called a *personal profit demand*), which is a comprehensive look at what you need, and then move on to what you want.

The following needs assessment (or personal profit demand) was taken from the pages of a shop management workbook with a 1979 copyright made available to shop owners by "Motor Age Magazine." Look at the expense amounts!

Table 5-1 Personal Profit Demand

Living Expenses (monthly)		
Housing (rent or mortgage)	$225.00	
Housing maintenance	$15.00	
Housing utilities	$45.00	
Food	$160.00	
Personal needs	$20.00	
Clothing	$25.00	
Life & other insurances	$30.00	
Transportation	$40.00	
Medical	$15.00	
Education/recreation	$35.00	
Gifts	$10.00	
Saving/loans	$45.00	
Miscellaneous	$150.00	
Total Living Expense	**$815.00**	**$815.00**
Business Loan	$5,000.00	
Monthly payments	$175.00	**$175.00**
Estimated Tax Liability	$3,600.00	
(Federal State, Local, and FICA) Annual	$300.00	**$300.00**
Total Monthly Personal Profit Demand		$1,290.00
Total Annual Personal Profit Demand (x12)		$15,480.00

While the amounts are almost laughable by today's standards, the categories and their respective expenses remain appropriate. All you have to do to make this exercise relevant for your world is to supply a more contemporary and realistic set of numbers (fill in the blanks in Table 5-2). Once you have done that, the only thing left is crunching the numbers.

Table 5-2 Personal Profit Demand

Living Expenses (monthly)		
Housing (rent or mortgage)	$	
Housing maintenance	$	
Housing utilities	$	
Food	$	
Personal needs	$	
Clothing	$	
Life & other insurances	$	
Transportation	$	
Medical	$	
Education/recreation	$	
Gifts	$	
Saving/loans	$	
Miscellaneous	$	
Total Living Expense	$	$
Business Loan	$	
Monthly payments	$	$
Estimated Tax Liability	$	
(Federal State, Local, and FICA) Annual	$	$
Total Monthly Personal Profit Demand		$
Total Annual Personal Profit Demand (x12)		$

THE NUMBERS WORK...

...if you work the numbers. Let's say you have decided that you deserve to make $10,000 per month, or $120,000 per year. It doesn't matter if the amount you chose is higher or lower. The amount is actually irrelevant. All that really matters is that you understand the principles involved in what we are about to do and why we are doing it.

In a conventional shop without a plan, whatever is left over after everyone else gets paid is what the shop owner gets to keep. There isn't much difference between that and throwing a fist full of twenty-dollar bills into the air and accepting the fact that any bills that remain suspended in the air are yours!

Getting Started

Let's start with the only thing we know for sure—we know we want to take $120,000 out of the business per year. What are we going to have to do to accomplish that?

What we are going to do is work backwards from that number. We'll place that $120,000 in the box at the top of our spreadsheet, as seen in Table 5-3. Then, we'll use everything we've learned so far to work the numbers. To do this exercise, we are either going to require a number of facts based upon actual history or we are going to have to start with some basic assumptions about our business, like our daily operating expenses without technician's pay, the number of days we are open, our parts purchases and average gross profit percentage on parts, as well as the number of technicians we employee, the number of technician hours we are open, our productivity percentage, and what those technicians cost us as a percentage of our labor sales.

Table 5-3 Reverse Budgeting

$120,000.00	Desired Profit
$1,047.23	Daily operating Expenses w/o tech pay
260	Number of days open
$202,666.00	Parts Purchases
36	Percent of Profit on Part/Accessories
33	Labor Cost as a Percentage
8,320	Number of 'Tech Hours' open
70	Percent Productivity
4.0	Number of Technicians
$120,000	Profit
$272,280	Expenses w/o tech pay
$202,666	Part cost
$594,946	Total outflow
$316,666	Part sales @ 36%
$278,280	Pabor profit needed
$415,344	Labor sales
$71.32	Labor sales per hour
$17.83	Labor $ per hrs per technician
$732,009	Annual sales
$61,001	Monthly sales
$15,250	Monthly sales per tech
$704	Daily sales per tech

If you own and operate a shop, your accountant or bookkeeper should be able to help you squeeze these numbers out of your financials. If you are planning to go into business and have no financial history to benefit from, use the target numbers we've plugged in here or the numbers we've been using throughout the series. Using the target numbers we've been using really can't hurt. In fact, it just may give you a higher goal to shoot for.

If you look at the illustration in Table 5-2, you'll see exactly where to plug those numbers in. Daily operating expenses without technician pay is $1,047.23. That is the sum total of all of your expenses with the exception of the cost of direct labor divided by the number of days open (the number of working days the shop is open for business). And, while that number may look and sound like a lot in our example, it's not. It's inexpensive compared to industry averages for shops with four technicians. This shop, Average Automotive, Inc., is open 260 days a year: that's five days a week, 52 weeks a year. Remember; if you are open Saturdays and/or Sundays (or if you plan to be…) be sure to adjust that number until it makes sense to you.

As discussed in two other books in this series, *Operational Excellence* and *The High Performance Shop*, our goal is to have our cost of direct labor run from about 30 percent of unloaded labor sales (without benefits and taxes) to 40 percent of loaded labor sales (with benefits and taxes). The *average* figure we have used here is 33 percent, which probably indicates the labor figure is unloaded. We'd also like to see an overall gross profit on parts of over 40 percent. Realistically, 45 to 55 percent would be more appropriate. If that number sounds *unrealistically* high to you, remember that you can margin up the parts that cost you less, within a certain price range, because most consumers are less price-sensitive to parts that fall within the lower price ranges. However, those same consumers are more likely to balk at higher margins accompanying high ticket items like rebuilt engines, transmissions, and cylinder heads, etc.

Regardless, there are still countless shops that show an overall gross profit on parts well below 40 percent, so we'll use 36 percent as our actual number and 40 percent as something to strive for in the future. You'll find all of these numbers in Table 5-1.

We can easily figure out our cost of goods sold—the cost of the parts and accessories we've purchased for resale—just as we can easily determine our service bay or technician productivity. The rest is just simple math.

Our *Expenses without technician pay* is the result of multiplying our daily operating expenses without technician pay and the number of days open: $1,047.23 x 260 = $272,280.

Our parts cost is the same number on both the top and bottom halves of the spreadsheet. Our Total Outflow or Total Expense is equal to expenses without tech pay ($272,280) + parts cost ($202,666) + desired profit ($120,000), which equals $594,946.

If we know what our parts cost is, and we know what kind of a gross margin we need in order to be successful, we know what our target parts sales should be. In this case, our parts cost us $202,666 and we are looking for a 45 to 55 percent GPM on parts. Realistically, however, we've never seen parts margins average higher than 33 percent. So, we'll shoot for a GPM on parts of 36 percent for now. To calculate what those target sales should be, divide your cost ($202,666) by the difference between the desired margin and 100percent. In other words, if our desired GPM is 36 percent, our parts cost would be 64 percent—$202,666 ÷ .64 = $316,666.

That number gets plugged in as your parts sales at 36 percent. At this point, you are ready to calculate the required dollar amount of your labor sales by subtracting the amount of your projected parts sales from your total outflow or expense. In this case, that would leave $278,280 in needed labor profit. Because you already know that your cost of direct labor is 33 percent of your total labor sales, you also know that your gross profit margin on parts is 67 percent. And because we've already figured out that all you have to do is divide the cost of direct labor by the gross profit margin on labor (which in this case is 67 percent) to get to the number you need—your projected gross labor sales—you can quickly figure out that your targeted labor sales will be $415,344.

Working backwards from the $120,000 profit that we are demanding from our business, we have a plan, or at least a target! We need to do $415,344 in labor sales and $316,666 in parts sales to achieve total sales of $732,009… with a gross profit margin on parts of 36 percent, a GPM on labor of 67 percent, and a parts/labor mix of 56/44—or 56 percent labor sales and 44 percent parts sales–and an overall gross profit margin of just about 45 percent (not really high enough, but OK for now!).

How closely do you want to monitor progress toward your goal—by the hour, week or month? If you are going to monitor labor sales by the hour, divide the necessary labor sales by the hours available for sale. In our model, that would be $415,344 divided by the total number of available_ hours (8,320 tech hours: 4 technicians x 8 hours per day x 260 working days per year), which would equal $49.92… but, that's only if you sold every hour you had to available to sell! Can you? Will you? Probably not! So, you factor the necessary labor sales by your productivity.

Now, do you really want to know how important productivity is? Let me show you!

At 100 percent, you need to sell $49.92 worth of labor per hour. With a projected productivity of 70 percent—still well about the national average for our industry—you have to sell $71.32 per billable hour (Remember, at 70 percent productivity you're only billing just over two of every three hours you have available to bill!) or $17.83 per technician per hour (four technicians). At 55 percent productivity, you would have to recover $90.77 worth of labor per hour or $22.69 per tech!

To make all this work, you would have to average $61,000 in sales per month, or $15,250 in parts and labor sales per tech per month—$704 per tech per day.

The purpose of these exercises is to help you understand that you don't have to be satisfied with whatever is left over. You don't have to pray for a couple of twenty-dollar bills to remain suspended in the air to know they're yours. You can establish a financial goal and then create a pro forma that lets you know exactly what the target looks like and focus everyone's attention on moving closer to that target. Without that kind of critical focus you will find yourself lost and struggling very quickly. With it, you will have a much better than average chance of actually achieving your financial target. And, if you don't… well, you'll get far closer than you might otherwise have gotten without something in your sights, without something to aim at.

The mantra of this book is *If you can't measure it, you can't manage it.* I think it is safe to say that we have just shown that what you and I do for a living can not only be measured, it can be managed, molded, and transformed into the kind of future most of us only dream about. All you and I have to do is paraphrase the old Nike television commercials and *Just do it!* So, what gets in the way?

SUMMARY

How much do you really need? How much profit is required to sustain both your business and your personal life? Take the time to consider the expenses in the personal profit demand worksheet and any expenses that are pertinent but may not appear there. Write them down and then add something for the risk and the responsibility involved in entrepreneurship. Factor in a reasonable ROI and write that information down. Then, take that total and insert it into spreadsheet we have developed for reverse budgeting and see what it will take to move from survival to success.

Use the chart on the next page to work the math issues before inserting them into the spreadsheet referred to in the Summary.

Thoughts

Actions

Results

CHAPTER 6

A Plan for Your Business

INTRODUCTION

This chapter will answer the critical questions you need to ask in order to establish a *strategic plan.*

It will take a look at that plan in terms of:

- Goals
- Strategies
- Objectives
- Tactics

It will teach you how to:

- Use the planned result as a tool to facilitate change
- Create an action plan
- Think out of the box
- Develop objectives
- Brainstorm
- Focus

QUESTIONS AND ANSWERS

Question: When do most business owners go to the bank or to a private lender for financial assistance?

Answer: When they are desperate and there is no alternative!

Question: What kind of information do they have available to help facilitate this request for financial aid?

Answer: Most of the time we'll never know because they don't know!

Question: What kind of information do they need?

Answer: Whatever it would take to make a compelling argument, like: How much money do you need? What will you use it for? How will you be able to pay it back? How long will it take? What kind of a return will this investment bring, both for you and for us as a lender or investor?

Question: How can you tell who has a business plan or a plan for their business and who does not?

Answer: By who gets the money!

You may think some of these questions and answers are a bit *tongue in cheek*, but they're not. Not really. Most business-owners are rarely prepared to speak to their banker in a language the banker understands and appreciates, let alone someone they don't know or someone unfamiliar with the business or their industry.

Consequently, I think it would be wise, if not at least prudent, to spend a few minutes talking about what is generically referred to in business as a *business plan*. In *Managing Dollars With Sense*—another book in this series—we'll be spending more time on this subject. But, we'll be approaching the business plan more from a financial perspective than a philosophical or practical one. For now, however, we will be concentrating on building a simple, yet practical business plan.

There are a number of different ways we can approach business planning. They range from the most comprehensive and detailed plans, to plans that are simple, general, and yet still practical. You can find volumes written on the subject, complete with countless templates that require just *plugging in* your numbers.

We will, however, be discussing a number of approaches that are consistent with my criteria for implementation in the automotive shop environment. That means they have to be compatible with everything else we've discussed so far and they have to be practical and easy as well. But, most of all, they must prove effective.

The 15 Question Business Plan

I would suggest starting with fifteen sheets of paper for the fifteen questions I am about to share with you—and since this is a plan for your business—fifteen sheets of paper for all the important individuals working inside or outside your business with a stake in your success. One question goes at the top of each sheet of paper, and is copied until there is a complete set for everyone involved. The questions I will submit here can be changed or modified to suit your particular needs, but I would suggest starting the exercise with these questions first just to get an idea of where the answers will take us.

Answering these fifteen questions will afford you the foundation you need to start building a plan that not only makes sense to you, but it will also help an interested third party understand who you are, what you are all about, and where you are taking your business.

What business are we in?

My hope is that the answer will vary depending upon which of the other books in our series you've read first. Realistically, most automotive service professionals would say they are a part of the automotive service industry—in the business of *servicing, maintaining, and repairing passenger cars and trucks*. While that is certainly an accurate answer, I'm not sure it is an adequate answer. In today's highly chaotic and competitive environment—an environment in which consumers have never been more knowledgeable or less forgiving—servicing, maintaining and repairing passenger cars and trucks just isn't going to be enough to insure your success.

Fixing the car right the first time is just what it takes to stay in the game!

It will take meticulous attention to every detail involved in meeting or exceeding the customer's wants, needs, and expectations consistently and without fail each and every time you are asked to perform. It will take having the vehicle ready when promised—not every once in a while, not most of the time, but just about every time. It will take being there when the customer needs you to be there, and it will take an unparalleled quality of both parts and service.

So the answer to the first question would certainly have to include the fact that you are in the automotive service industry, servicing, maintaining, and repairing your customer's cars and trucks. It would also have to include a statement of purpose consistent with meeting or exceeding customer wants, needs, and expectations as well… something like:

> Schneider's Automotive Repair is in the business of insuring the personal freedom and mobility of our customers through the care, maintenance, and repair of their passenger cars and trucks. We do this by understanding and anticipating their wants, needs, and expectations almost before they do… and then by meeting or exceeding those wants, needs, and expectations each and every time service is delivered.

Who are our customers?

Again, the answer to this second question might be different depending upon where you find yourself on this journey. If you understand how important it is to choose the right target market for your products and services, you have already come to realize that you can't work for everyone because not everyone will appreciate the level or quality of the products and services you provide.

You really have to take the time to determine the level of quality and service you will demand from yourself and from your associates and then profile the kind of customer who will respond. Or, you can survey the demographics of the neighborhood you will be serving and then profile a target market within that neighborhood—focusing on the kinds of products and services they both want and need. Either way, you'll have to describe your target customers (more than one target) in great detail.

One of our targets is:

> Single women, heads of households, gross income of $35,000 or more…

Our description is really quite a bit more detailed than just that, and includes a number of additional qualifiers like the makes and models of the cars, trucks, and SUVs they drive and that we prefer to work on. Another target is:

> "Boomer Geezers," (Boomer Geezers is a term used to describe the 76,000,000+ Americans who are now 55-years old and older) three drivers or more in the family, gross family income of $100,000 or more...

Either example should give you some idea of how you can qualify—and then quantify—the kind of people you would like to invite to experience the products and services your business has to offer. But in the end, try to find targets that fit the profiles of those customers who make you smile every time they pull down the driveway. That way you'll always feel good about the commitment and effort required to deliver world-class service to all your customers all the time.

What do they want, need, and expect from us?

In any discussion of customer wants, needs, and expectations, we cannot escape the obvious: *Fix it right the first time!* But in every customer satisfaction survey we've seen over the past few years, that is not the number one customer concern. The number one issue for the customer is: *Ready when promised!* Having the vehicle available when the service facility said the job would be completed is critical to lasting customer satisfaction. Having a plan that ignores something that critical would be incomplete at best—dangerous and misleading at worst.

Time is perhaps the most critical commodity of our era. Everyone wants more time—no one has enough time. The demands of work, friends, and family have each of us stretched to the limit and our failure to remain sensitive to that reality leaves the customer no choice but to think we don't care.

So, in answering a question like the one above, I would start with the consumer's need to have the vehicle *ready when promised*, and then go on to *fix it right the first time*, following those two with explanations of meeting customer wants, needs, and expectations for convenience, quality, accessibility and response-ability—the ability to respond quickly and completely to the customer's demands.

How do we make our products and services available?

This question goes to the heart of meeting or exceeding customer wants, needs, and expectations by understanding when and how they—our customers—want access to those products and services. It has to do with staffing and hours of operation—with before or after-hours vehicle pickup or delivery—and it has to do with how we package those services.

Do you plan on being open in the evenings, on Saturdays, or even on Sundays? If it was up to most vehicle owners, you or someone like you would be at the shop 24-7-365. But, barring that, when will you be open? Will you be open early enough to catch your customers before they have to leave for the office? And when will you close? Will you close before they have a chance to stop by, pick the vehicle up, and pay you?

These are all questions that can and should be addressed in the kind of plan we are discussing here.

Who is our competition?

In the midst of a heated discussion centered on neighborhood competition, a good friend of mine taught me an invaluable lesson. He was the consummate businessman, and as I complained endlessly about what my competitors were either doing or not doing, he quietly said: "I have no competition."

At first, I was really amazed. I know his business and I know how and where he goes to market and believe me, it is not without its share of competition. But when I pressed him for an explanation, his meaning became clear:

> *I'm the competition! I'm the one everyone has to watch out for! I'm the innovator! I'm the one committed to understanding my customer's wants, needs, and expectations better than anyone else in my market. I'm the leader. They're the followers. By the time they have imitated something I did yesterday, I've already evaluated whether or not it worked and gone on to something else. I'm always moving forward, while they're always at least one step behind and trying to catch up! Since, I'm out in front by myself, I have no competition!*

That one paragraph changed my business life and I can now honestly say that I run my business as if I have no competition as well. However, for the sake of a business plan, identifying groups of competitors is not only prudent—it's vital. No one will take your plan seriously if it does not include at least a list of your competitors and a description of who they are, as well as an explanation of what they do. My recommendation is to place other service facilities in your market area in *tiers*—first, second, third, etc.

First tier competitors would be those service facilities that provide their customers with the same *quality* of service you do. The easiest way to identify such competitors would be to ask yourself who you would send one of your favorite customers to if you couldn't work on the vehicle yourself.

My guess is there are at least two or three service facilities in your general area that would fit that criterion and this first tier could also include a number of factory service outlets.

The second tier could be comprised of specialty shops, automotive franchises, retailers, and mass marketers. The third tier could be the quick service facilities, and so on.

What advantages do we have over our competition?

This is a really important question to both ask and answer. It will help you understand exactly what your service capabilities and capacities are when compared with other service providers in your market area. Answers would range from location to technical capability, from hours of operation to number of service bays. But whatever the answer, it should call for a fairly extensive assessment, and consequently an honest evaluation of just where your products and services are superior.

What advantages do our competitors have over us?

Just like the answers to the question above, the answers to this question require careful assessment and evaluation. However, unlike the question above, it is more difficult to answer honestly because it forces you to confront your own failures and/or weaknesses. It forces you to look at where your competition is really capable of using adjectives like *more* and *better*. But it is through this process that you will uncover the areas you need to confront in order to achieve a leadership position in the marketplace.

What do our competitors do that we don't do?

Once again, we need to look at things that just might make us a little uncomfortable—like whether or not our competitors are open when we are not, or whether they have made an investment in training, equipment, or technology that allows them to perform in areas that we are not able to serve. It almost demands that you develop a service menu for your facility and then a comparative list for your competitors. By the process of evaluation, you will not only have the answer to this question, you will have the answers to the next question as well.

What do we do that our competitors don't do?

The answer to this question will materialize as we answer the question above. In essence, it is the things we are capable of doing and doing well that our competitors can't or won't do. The answer could be a specific service or a whole area of services. Whatever it is, it will point the way toward a competitive edge.

Who are the owners of this business and what are their roles and strengths?

In too many instances, this may be the first time this question has ever been asked and answered. A clear description of the owner's responsibilities, roles, and strengths is critical for success—especially when there is someone in a secondary management position, such as a general manager or a service manager. Without such a description, responsibilities become cloudy and things—sometimes very important things—fall through the cracks.

Perhaps the best example of all this is what we experienced in our own business. There are four Schneiders at Schneider's Automotive. It wasn't that long ago that we were stepping all over each other's responsibilities—doing each other's jobs. Ultimately, the chaos created by never knowing if or when something got done, or if more than one of us had done the same thing, got to be too much and we finally had to split things up into areas of responsibility with formal job descriptions.

Who are the managers of this business and what are their responsibilities?

The same person who taught me about having no competition taught me another lesson as well. He taught me that you *hire your weaknesses*!

The concept is simple. We all like to be around people like us! However, in business, you don't need to hire someone to do something that you are really good at rather than doing it yourself—unless you no longer want to do it. It is more important—critical, in fact—to hire individuals capable of doing well that which you are not good at doing at all. If you are not a *detail-oriented person*, you need to hire someone who is meticulous in everything they do—someone obsessed with detail. If you're the one who is detailed and you find yourself reluctant to get close to your customers, you need to hire a *people person* to do that job.

Regardless, to do that you have to understand and know your strengths and weaknesses and staff your management positions with those individuals who are strong where you are weak.

What financial information do we have available and when?

I have actually been in shops when the financial statements arrived and watched as the owner/principle opened the manila envelope, removed the financial statements, looked to insure the number at the bottom of the page wasn't in brackets, and then slid the financials back into the envelope—never to be looked at again!

That is not financial analysis. It's financial suicide!

This information is critical, and because financial information is really *historical* in nature—it is derived from information concerning events that have passed—the sooner you get that information the better. In other words, you need everything you can get, and you need it as fast as you can get it. Time is really of the essence in every sense of the expression.

What financial information do we need?
We need a full array of financial statements including—but not limited to—a current *income statement* or *statement of accounts*, a current *balance sheet*, a *cash flow analysis*, and a *break-even report*. You need an accurate and current listing of your *account receivables*—money that is owed you. You also need an accurate and current listing of your *accounts payable*—money that you owe. You also need a current and accurate *inventory report*.

What changes will take place in our type of business over the next two or three years?
It would be wonderful if we could answer a question like this, but in all honesty we probably can't. We can, however, project what we do know into the future. For instance, we know that the explosion of technology we have been confronted with over the past twenty years is not going to stop. We know we will be looking at hybrid vehicles, 42-volt electrical systems, and (ultimately) fuel cells. We can be pretty sure that the shortage of trained and qualified technicians we've been experiencing isn't going to somehow *fix* itself. And we know that the competition for automotive service dollars will become more intense as well.

If we know all that, we can begin to think about how we will respond to those challenges, and that is what the planning process is all about.

What are our company's current problems and/or desired goals?
By taking a long, hard look at our business—its capacity and capability to deliver world class service consistently—we have gone one step beyond the majority of other service providers working in our industry. By identifying our problems and listing our opportunities—by itemizing the threats and considering both or strengths and our weaknesses—we have demonstrated that we are aware of the environment and conditions in which we perform and have both the desire and the ability to succeed.

Specific Elements of A Plan

A formal business plan should include both a statement of the company's vision as well as its mission statement. As we have discussed, that mission statement should clearly describe what you perceive to be the company's long-term mission.

It should also include professional biographies for each of the individuals involved in managing the business, as well as biographies for key employees and the owner or owners. These biographies should encompass where they have been and what they have done while involved in the industry they serve, as well as the number of years they have been involved.

It wouldn't be a bad idea to include an explanation and history of the automotive service aftermarket—its history, the industry as it exists today, and projections for the future. This should address an analysis of the marketplace, who the key players are, how they fall into your competitive matrix, and an explanation of how and why you will not only survive, but succeed. In doing that, it should address the company's strengths, weaknesses, market opportunities, and threats.

It would be wise to underscore your understanding of the marketplace by identifying certain consumer problems, along with an explanation of how you plan to exploit the opportunities created by those problems. All of this should be wrapped around your philosophy of doing business—or your business concept, as it is commonly called.

Identify the various tiers of competition as described earlier and explain how and why you feel you have the competitive edge.

Enumerate your company's goals, objectives, and tactics—what you are going to do and when you're going to do it—in order to succeed. These can and should be broken down into long-term (3-5 years) and short-term (within 1 year) goals and the specific objectives associated with each.

State your specific goals for *customer share* and *market share* and how you intend to achieve those objectives.

Identify your specific financial goals, particularly in reference to your KPIs and show how achieving those objectives will greatly enhance your overall financial picture.

Discuss what it's going to take to make all this happen—the resources, technology, and staffing necessary—and then talk about risks and rewards.

SUMMARY

Do you have a plan for your business? Is it written, or is it just a *loose* sense of what has to be done that bubbles to the top of your consciousness every once in a while? Take a moment to think about those things most critical to your success or the success of your business. Think of the things you would like to do or have to do to go to the next level of performance.

Write the answers to these questions down in the chart on the next page and then write down the one or two things that would have to happen for each to become a reality.

Thoughts

Actions

Results

CHAPTER 7

Planning Strategically

INTRODUCTION

This chapter will help identify the key questions to ask when formulating a *strategic business plan.*

You will learn:

- The principles of *scenario planning*
- How to redefine goals, strategies, objectives, and tactics
- How to use *the planned result* as a tool
- How to develop strategies
- A process for *thinking outside of the box*

MBWA OR MBLOTW

Our advice flies in the face of conventional work ethic wisdom. Based on experience and observation, we recommend spending at least one full day each month out of the office doing nothing but working on your business.

Jason Jennings and Laurence Haughton
It's Not The Big That Eat The Small... It's The Fast That Eat The Slow

MBWA: Managing by Wandering Around

MBWA first appeared in the landmark business book, *In Search of Excellence*, by Tom Peters and Robert Waterman. "Management by wandering around," they wrote, "may be the most important thing managers can do to improve work quality and productivity."

MBLOTW: Management by Looking Out the Window

MBLOTW is mine and it shares the same important message that Stan Stephenson, former Editor-in-Chief of Motor Age Magazine, first articulated more years ago than I can remember, and that is: "You can't manage your business from a creeper underneath a car."

What we're really talking about here is keeping it real—keeping things in perspective. We're also talking about working on your business and not just in it! That's what everything we've been discussing has been all about.

If you don't take a few minutes out of each day to wander around, to scope out your competitors, or just to stare out the window and wonder about your future—the future of your business or the future of your industry—who will?

OTHER QUESTIONS TO ASK

What is the real potential for this business?

What will the market—your specific corner of the market—really support?

Realistically, this is a question of demographics. Are there enough people or families in your marketing area that meet your target demographic to support your business in the future? Are there any demographic trends in your neighborhood or in the surrounding neighborhoods you should be aware of (i.e., people moving in, families moving out)?

If you dream about opening up a shop specializing in high-end European cars and SUVs, you should be looking for areas with high-density, high-end European vehicle registrations. You should spend a minute or two thinking about whether the majority of those vehicles are leased or owned, as well, because it can make a big difference in not only how well they are maintained, but if they will be maintained at all!

Do you need to be offering other products/services to achieve your full potential?

Are there potential profit centers you are missing? Would offering other products and services take your eye off your target market—your target customer? Or would they add profit dollars to your bottom line? What are the potential rewards for entering a different service arena? What are the risks? Planning is all about analyzing situations and the accompanying potential for risk or reward.

What other products and services could you offer? How crowded are those market segments in your area? Is there room for another player? Or is there room to create a whole new category? Can you create a totally new and different service—a totally new solution to a completely different problem?

What is your really BIG objective?

Some companies call it a BHAG—a *Big-Hairy-Audacious-Goal*—something so outrageous, so huge, it becomes utterly transformational! What kind of a BHAG could you create to revolutionize the way automotive service is delivered in your community? The Harvard Business School used to say: *Solve a problem—start a business!*

How quickly could this objective be achieved if everyone in the company turned all their efforts toward achieving this objective? What would it take to make that happen?

What are you doing and why?

Is there anything that you are doing today, anything that is an integral part of your daily operation that you can't explain? Are you doing the same thing, the same way, every time you do it—waiting for the results to change somehow? Are you doing the same things every day without a clue as to why you're doing them or how doing them that way started in the first place? Can you find anything in your daily regimen that can or should be destroyed or abandoned?

What additional human and financial resources would be required to achieve your big objective?

Have you quantified them? Do you know exactly what you would need? Do you have access to those resources?

We're in the middle of one of the most acute shortages of technical help the industry has ever seen. Wanting more or better technicians, and being able to find them, are two entirely different realities! Perhaps it would be more appropriate to contemplate what we need to do to keep the technicians we already have so nobody steals them from us?

What would you do with the wealth created?

How badly do you really want to achieve that big objective? What are you willing to invest in the way of time, money, or other resources to achieve it? How quickly could this big objective be realized?

Do you have the right people in the right positions to achieve it? How many more customers would you need and what would each new customer have to spend in order to reach that big objective?

If you were successful, what would you do with the money?

Is the world changing faster on the outside than you are changing on the inside?

Do you have departments or products or services that should be *dumped*? Are you strong enough to make those hard decisions and then hard enough to implement them?

SCENARIO PLANNING

Only the paranoid survive!

Andy Grove, CEO of Intel

Scenario planning is really nothing more or less than calculating the *What if's...* before spending a lot of time on the *How to's...* It's not a bad idea. Nothing will sink a *How to* faster than a *What if* if you didn't plan for it properly.

Putting it another way, scenario planning is a process for guiding *forward thinking*. You ask questions that begin with *What if...* and then create tactical and strategic responses. It should involve all the key people in your business.

The enemy of scenario planning—or any other kind of planning for that matter—is time. Why? Because there just isn't enough time to do all the important things you have to get done right now and still plan for the future. Gary Hamel, author of *Competing for the Future*, suggests that most corporate executives (that includes YOU!) spend less than 2.4 percent of their time thinking about the future. Based upon a 60-hour work week, that's less than 20 minutes per day...

You tell me... Is that enough?

Institutionalize Planning

The only way to insure there is enough time dedicated to the planning process is to formally schedule planning as a corporate or company function. In this case, you could schedule a regular meeting with scenario planning as its primary focus.

Include everyone with a stake in the company's future—that includes just about everyone involved in the company from you as the principal shareholder or manager, to the person responsible for cleaning the restrooms. It might not be a bad idea to include certain key suppliers as well. They just might be able to help provide insight and perspective otherwise unavailable to you.

Don't overlook non-participants either. Just because your wife or husband isn't intimately involved doesn't mean they don't have something to offer. And be sure to allow enough time for discussion.

Begin by suggesting a plausible future possibility—like the emergence of a new competitor, the introduction of new technology, a dramatic change in the marketplace, or a compelling new development or trend.

Record everything! Keep a running record of all recommendations and suggested actions.

Monitor what you decide to do and your results against actual changes in your current reality. After a while, you will be surprised at how good you can get at predicting the future. But more important, you will be surprised at how well prepared you can become when it comes to meeting the challenges a new and different future will bring with it.

Institutionalize New Ideas

Often, one of the positive benefits of scenario planning is (or at least can be) a new process to help you respond to the challenges of running a business.

> **proc·ess** (prä'ses, prä'səs) *noun Abbr.* proc. **1.** A series of actions, changes, or functions bringing about a result. **2.** A series of operations performed in the making or treatment of a product

Each process becomes a part of a system.

> **sys·tem** (sis'təm) *noun Abbr.* syst. **1.** A group of interacting, interrelated, or interdependent elements forming a complex whole. **2.** A functionally related group of elements. **3.** An organized set of interrelated ideas or principles. **4.** A social, economic, or political organizational form. **5.** A condition of harmonious, orderly interaction. **6.** An organized and coordinated method; a procedure.

When you institutionalize a *process* it becomes a *practice.*

> **prac·tice** (prak'tis) *verb* **1.** To do or perform habitually or customarily; make a habit of. **2.** To do or perform (something) repeatedly in order to acquire or polish a skill

In the world of planning—especially the kind of planning involved in actively engaging yourself and your associates in the future—scenario planning is one of the most important exercises you can undertake in order to transform your business from *reactive* to *proactive.*

Zones

It is so important that a company like Charles Schwab has institutionalized scenario planning to the point it carries its own copyright. Schwab THINK (THe Innovation NetworK©), as it is called, operates in four zones or areas of corporate specialization.

The first of the four zones is *Idea Central.* In Idea Central, ideas are accepted from everyone and anyone willing to suggest and submit.

The second of the four zones is *The LOOP.* The LOOP is tied to Schwab's web presence and gathers its ideas and suggestions from the website.

SMART is the third zone and it is used to gather suggestions and ideas from and for the telephone customer care centers.

The fourth and final zone, at least for now, is *Venture Quest*. Venture Quest is for big, revolutionary ideas requiring a significant capital investment.

Schwab's guideline for moving forward with any of the suggestions submitted to any of the four zones is fairly straightforward. Does it fit our guiding principles (see Chapter 4, page 47)? Which strategic priority does the idea support the most? Who will benefit most? What specific need or opportunity does the idea tackle? What is the problem being tackled? What are the success factors? What set of rules would the innovation break? (Schwab has found that their most successful ideas have resulted from breaking their own rules.) Who would your idea upset the most? How destructive would it be? Finally, who supports the idea?

A SELFISH MISTRESS

You should remember that a business, your business, any business, can easily become a selfish mistress!

It will take everything you have to give it and ask for more without the slightest bit of shame or remorse. In fact, it is quite capable of taking far more than you realistically have to give! But, by the same token, it can be incredibly generous if you can learn how to coax that business into performing for you instead of the other way around. The key is realistic and reasonable expectations, so let's take a look at this for just a moment, as we get ready to move on.

Desired Outcome

In the last chapter, we looked at *personal profit demands*, *needs assessments* and *reverse budgeting*—all the tools necessary to work backwards, from the inside out, from the bottom to the top, from your physical needs to your financial wants. We're going to do the same thing with the other, non-financial aspects of your business.

Before we do that, let's take a moment to clarify what it is we really want from our business, for our lives and for our families.

Are you looking for increased earnings, greater profits, reasonable growth or operational excellence? Are you looking for better technicians with a higher level of technical competency, less stress, a different lifestyle, financial security, more time to spend with the family, more time for quality recreation, more time to travel, or more time to just sit and relax?

Now would be a great time to write some of those dreams and desires down. They will become the heart of your strategic plan!

What Are You Looking For?

So, what is it you are looking for in your life and in your business? Which of the examples listed above would you choose? Why?

Which should enable the other? Should your business fuel your life or should you give everything you have to give to your business?

The answers to the questions asked above will become the foundation of your vision, the purpose of your mission. They will define your strategies, goals, objectives, and tactics.

Redefining the Terms

Anyone who has known me for a while, read my columns, or attended one of my seminars knows that I'm a bit neurotic when it comes to words, their usage, and particularly their definition. You've seen how that neurosis manifests itself to a small degree with regard to the boxes that appear throughout the text and the words and definitions within. Having shared thoughts and ideas—both in print and through presentations—for the better part of twenty years, I've come to realize that defining key words in terms of how they will be used is critical to achieving any kind of substantive communication.

Consequently, we'll be spending the next few minutes reviewing the terms we have been using—and will continue to use—throughout the rest of this chapter.

- **Goals** are something you hope to attain or achieve, it is something you work for or toward: the purpose toward which an endeavor is directed. Goals are the target—the objective. They determine what your strategic objectives will be.

goal (gōl) *noun* 1. The/an objective.

- **Strategies** are the broad programs or master projects that you will design to achieve specific goals
- **Objectives** are the specific and measurable results that will lead you toward the realization of your goals by moving you incrementally closer to your vision.
- **Tactics** are the smaller bits used to achieve objectives.
- **Strategic Planning** is deadline driven and time specific, because without a deadline goals are not goals. They are dreams. Strategic planning is a conscious and proactive effort to shape the destiny of an organization. It is a response to forces both internal and external that would take the everyday decision-making process out of your hands. It recognizes the fact that you will either determine what your company is all about, what it stands for and where it's going… or you will let outside influences or a lack of cohesion and focus inside your organization determine that future for you!

But the real question here has to be how much real planning is going on in our industry to begin with? Or more to the point, how much real planning are you doing?

Putting It All Together

Let's start putting it all together—starting with a long hard look at the ideal model of a business we keep in our heart or in our mind and then reconciling that with our current reality.

Do we have a philosophy that dictates who we are and what we do? Do we know what we are all about as far as being a company is concerned? If we do, does that philosophy match our *current reality*? Is that philosophy consistent with our mission, vision, goals, and objectives? Does that vision paint a clear picture of our enterprise in the future? Can this be stated in a concise declaration of why the business exists? Do we recognize that the ultimate purpose of our mission is to achieve that vision?

Will everything we do remain consistent with our values? Will it work in concert with our guiding principles? Will our business be built on our own unchanging, enduring beliefs? What we hold dear will act as a kind of moral compass, keeping us on track with regard to important character issues like honesty, integrity, and morality. This will become the guideposts of our behavior in everything we do.

To endure… to be successful… every relationship we enter into with the customer, with our suppliers, or with the people with whom we work will have to be based upon mutual respect and open communication, as that more than anything else will reflect the system of management we adopt.

If, for some reason, you aren't sure just exactly how important these key issues really are, all you need do is focus on the collapse of corporate giants like ENRON, WorldCom, Arthur Andersen or any one of a number of other companies who put short-term gain ahead of long-term stability and success. These are all prime examples of exactly what happens when executives and corporations lose their way—lose the moral compass that guiding principles can provide.

OUR MISSION AND VISION

We believe that our mission, vision, and guiding principles serve to keep us on track... to constantly remind us of what's important, who we serve, and what our responsibilities are.

> To reinvent Schneider's Automotive Repair into a Total Quality Service-oriented, customer-driven company focused on providing quality automotive service to our customers, improved quality of life for our employees, and increased profits to our stakeholders by doing what we do best...
>
> Providing the Needed Work, Done Well, Finished on Time and at a Fair Price, With All Work Guaranteed.
>
> Ultimately, becoming as important to our clients as they are to us!

That is our mission, our target, and our ultimate destination. If we are successful, accomplishing that mission moves us ever closer to making our vision a reality. What if we fail? Well, if we fail we will be closer to accomplishing that vision than we would have ever gotten without something to guide us—something that takes us clearly into the future and helps direct our decision-making.

For us, it came together in the following vision statement:

> Schneider's Automotive Repair, Inc., will be the premier supplier of high quality automotive products and services.
>
> Our future will transform our customer's reason for doing business with us from purchasing automotive parts and services to purchasing confidence and security.
>
> We will be recognized in our area and throughout our industry as knowing our customer's wants, needs and expectations as well or better than they do.
>
> We will be a partner to our customers and provide value-added products and service in anticipation of their needs.
>
> We will continually improve the automotive service experience of our clients, the profits of our corporation, the growth and welfare of our people, the image of our industry and the betterment of our community.

Having a vision and a mission like those above is the kind of solid foundation you will need for a future that is anything but predictable. It results in the kind of consistency that progress is built upon—a consistency that both your clients and your employees crave. It provides guidelines and mile markers to help chart a proper course. It can also become the ultimate dispute resolution tool when more than one person is responsible for the success of the enterprise. Things are either consistent with your vision and your mission or they are not. They are either relevant or they are extraneous, superfluous, or unnecessary.

You mission, vision, and guiding principles are focused on the future—your future—not on the present. And, they will determine your strategies, objectives, and tactics.

Remember that your objectives are achieved by employing tactics and using strategies. They are your highest priority outcome and induce change when realized.

Tactics are specific, smaller, time-phased activities that make it possible to achieve your objectives, as seen in Figure 7-1.

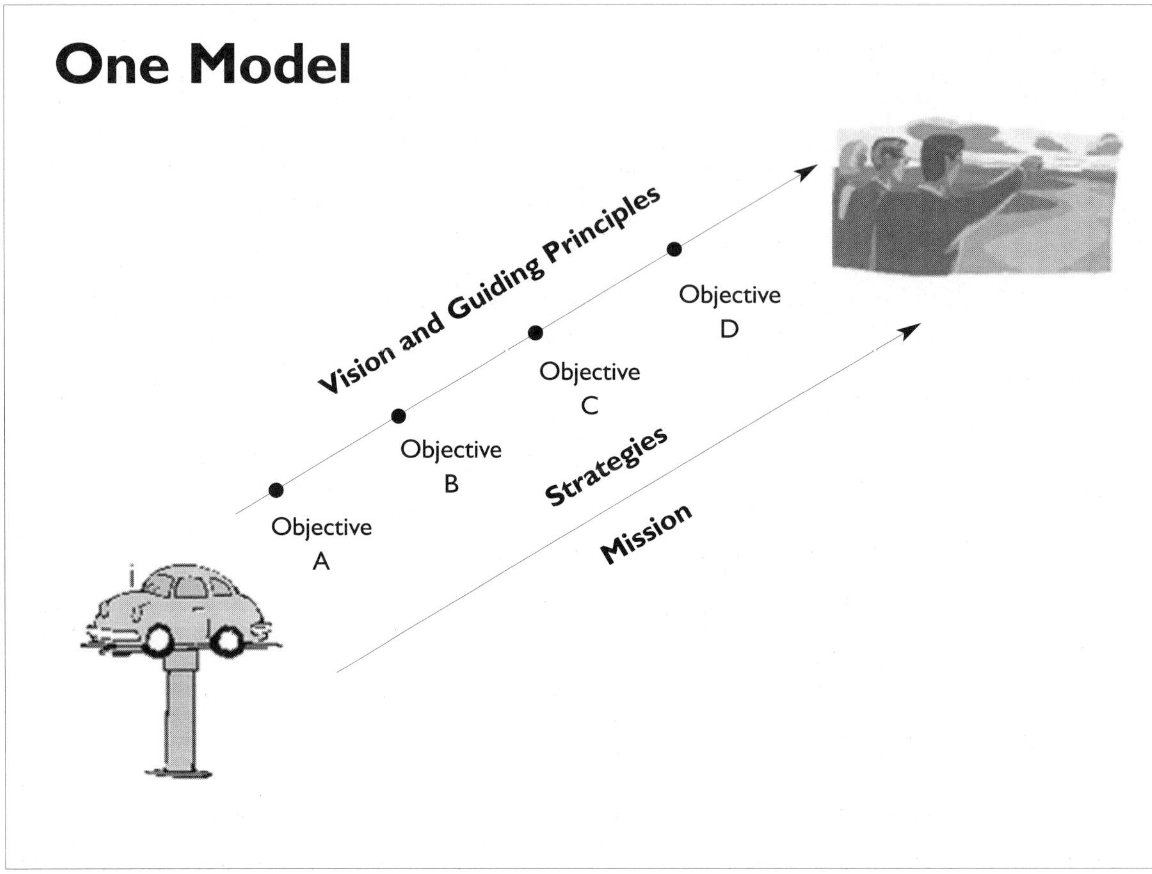

Figure 7-1 Have a vision and a mission.

The Planned Result

Clearly, one of the principles of reverse budgeting is to create a financial model of future income and then establish the financial goals and objectives necessary to make that happen. It is also possible to paint a very clear and vivid picture of what life would be like if we had already achieved our vision, and then work backward from that vision to determine what would have to happen for that vision to become a reality.

For our purposes, we can call that *the planned result* because, after all, that's exactly what it is. We can begin by looking into the future when our vision has become reality.

In that future

1. We fully understand and appreciate our client's wants, needs & expectations.
2. As a result, they come to us not only for automotive service and repair, but to insure their confidence and security as well.
3. In this future, we are recognized as the premier provider of automotive service in our market.
4. Our productivity is excellent—way above the national average.
5. We have met or exceeded all of our production goals and objectives.
6. Our relationships with both our customer and our employees are excellent.

It would be wonderful if the above statements were true in every case, but they're not. Not yet. We have been working diligently to create exactly that kind of a reality for everyone associated with our business, but we're not quite there yet. Despite the fact that they are not 100 percent accurate, they serve a very important purpose. You see by stating what the planned result would be, you can identify what it would take to move from where you are to where you want to be.

Action Plan: Strategies

Action plans and strategies look at the above statements with a critical eye, wondering what it might take to make that great leap from your current reality to the realization of your dreams. Let's look at them one at a time while we address the first bullet point as an example of how we can work our way through each of the remaining bullets.

For us to say that *we fully understand and appreciate our client's wants, needs, and expectations* we would have to

1. Increase the level of awareness, knowledge, and understanding of the corporate vision to everyone on staff.
2. In conjunction with that, we would have to establish a process to cultivate, harvest, and improve our knowledge and understanding of customer wants, needs, and expectations.
3. This would have to include developing the training and support necessary to promote all of our continuous improvement efforts throughout the company.
4. We would have to identify and exploit any and all strategic advantages we could.
5. And, certainly, we would have to find the wherewithal to substantially increase revenues in order to facilitate all of the above, while constantly monitoring our current reality to insure that our assumptions about ourselves and our world are still accurate!

By answering the question *What would it be like if all my dreams for the future came true?* we have identified a number of different areas that would require change. Once the level of that change can be quantified (the specific vision you would like to achieve), the metrics to monitor that change can be established. At that point, the only thing remaining would be to implement the various action steps required—monitoring your progress throughout the exercise.

The Challenge

We can't assume that the industry we know today will be the industry we will inhabit in the future. We must constantly invent new ways of exceeding customer expectations every day. We must reinvent ourselves to accommodate changes in technology, the culture, and in our customer base.

How To:

No one lives in a vacuum. Even in a one-man shop, with the level of support required for success today, it would be hard for anyone to honestly proclaim that they *work alone*.

Everyone is smarter than anyone! So, the first thing to do would be to assemble a team of all the important people in your life, or in your business, with a stake in your success. *Building A Team* is both the title and the subject of yet another book in this series. However, for now, we can say that this team may consist of just you if you are brave enough, or it could be made up of you and your wife—or you, your wife, and an employee, or two, or more. To be successful, you must involve all the important people involved in both your home and in your business life.

Think Out-of-the-Box

No one knows what the future will bring, especially a future transformed by your individual vision. In order to contemplate such a future, you almost have to suspend judgment. Since the future hasn't happened yet, no one can state with any degree of certainty what it might be like. Consequently, there really aren't going to be any *right* or *wrong* answers.

In order to create such a future, you are going to have to look hard at a number of *what if's*, recognizing that each one of them is equally valid no matter how outrageous or *unrealistic* they might be. Remember, it wasn't all that long ago that people were sure the world was flat, that man couldn't fly, that there was no need for more than 5,000 personal computers *in the world*, that there was no commercial application for fax machines, and that people would never shop online!

Assessment

Assessment is just another way of saying *Ask a lot of questions!* The only difference is there is a process of evaluation implied. What kind of questions? Questions like: What is the culture of this company—as seen by you, your employees, and your customers?

What are your *core competencies*? What are those things you are especially good at—those things that helped you build your reputation. If you aren't sure you have *core competencies*, answer this question: Do you ever receive telephone calls from other shops or other technicians asking for help in a certain area of technology? If you answer Yes! and these other shops and other technicians depend upon your knowledge and your expertise in a certain area of automotive service for their success, you have just established an area of core competency.

SWOT

One of the most powerful assessment tools I know of is the SWOT Exercise. SWOT identifies Strengths, Weaknesses, Opportunities and Threats.

Strengths and weaknesses are the assets and liabilities over which you should (or may) have some control. Opportunities and threats are conditions external to the organization and represent conditions that are not subject to your control.

Review

If you have a mission statement, review it. If you don't have a mission statement, write one! Remember, the key question will always be something like: Why does this company really exist?

If you have taken the time to consider what principles guide both your personal and corporate behavior and then written them down, review them. If there isn't such a written set of guiding principles, write them!

Ask yourself: What kind of a time frame are we working with? Long-term goals and objectives are generally assumed valid for three to five years. Short-term goals and objectives are those things you would like to achieve or accomplish within one year.

Establish a set of *assumptions* about your business. Assumptions are the key factors that will impact your competitive position during the time-frame established by your plan. What assumptions must you make about the economy, the proliferation of technology, the competition, the critical shortage of competent and qualified technicians, changes in the marketplace (distribution), or our image—both *public* and *self*?

Develop Objectives

What do you have to do to make your vision a reality? Do you have to access more technical training for both yourself and your employees, or will more and better management training meet your needs? Do you need to focus your attention on your customers with better customer relationship management training for you and your staff? Would your goals and objectives be better met by researching and ultimately purchasing new and different equipment? Should you be considering a different array of products and services or would it be better to completely overhaul the way service will be delivered in your facility?

Brainstorming

When you start generating ideas, you have to continually remind yourself that every idea is *legitimate*. Suspend judgment—especially your own judgment. No idea is *ridiculous*, so write everything down. Let one idea freely lead to others, generating as many ideas as possible in a fixed and predetermined amount of time.

Let ideas *incubate* for awhile. Periods of silence—even prolonged periods of silence—are okay.

Once you have allowed your list of ideas to percolate for a little bit, you can start a process of elimination. Realize that 20 percent of the ideas you have will provide 100 percent of the improvement necessary to fuel your efforts to achieve your vision—if they are the right twenty percent! The idea is to review all the ideas, with the express purpose of selecting the top ten or twenty percent as the most critical by voting on them.

The Process

There are a number of different ways to approach the selection process. We'll be looking at two of the most common and effective. One is referred to as the *Affinity Diagram*, and the other is the *Interrelationship Diagram.*

The Affinity Diagram is illustrated in Figure 7-2.

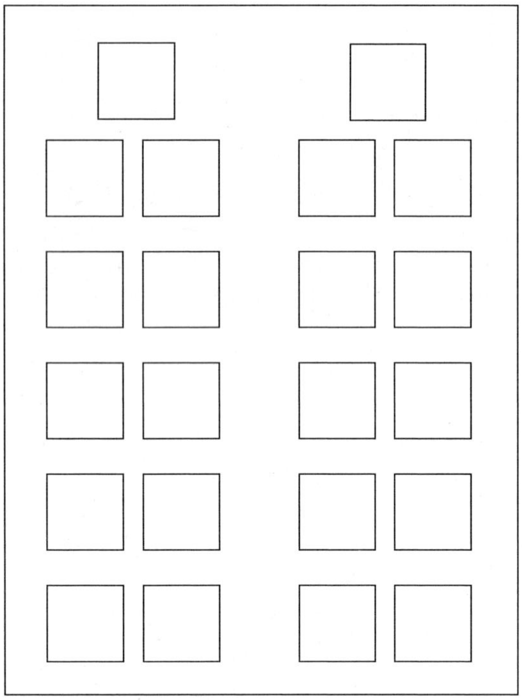

Figure 7-2 The Affinity Diagram (© 1998 Brenda Barnes & Jim Van Wormer)

The Affinity Diagram is the result of an interactive data gathering method that allows groups of people to identify and process large quantities of ideas in a very short period of time. It is a non-judgmental way to collect and process those ideas. It works as follows:

1. A question is posed.
2. Team members involved in their own private brainstorming sessions write down their thoughts and ideas separately on 3 x 5 index cards or sticky notes.
3. The team members then place them randomly in the middle of the table.
4. The ideas are then grouped into general categories of ideas, with a header placed at the top of each group for identification purposes.
5. After they are grouped, they are evaluated and assessed for substance and relevance.

Relations diagrams are drawn to show all the different relationships between factors, areas, or processes. They are worthwhile because they make it easy to pick out the driving factors in a situation. For example, a relations diagram of urban poverty might start out something like the one in Figure 7-3.

Like the Affinity Diagram, you start with a question or a premise and work from there. In the example above, we are looking at a situation that is called *urban flight,* which is the abandonment of our inner cities. The topic sparks a discussion focused on both the causes and the effects of the phenomenon in question. These are identified on 3 x 5 index cards or sticky notes and pinned to a board, or placed on a wall.

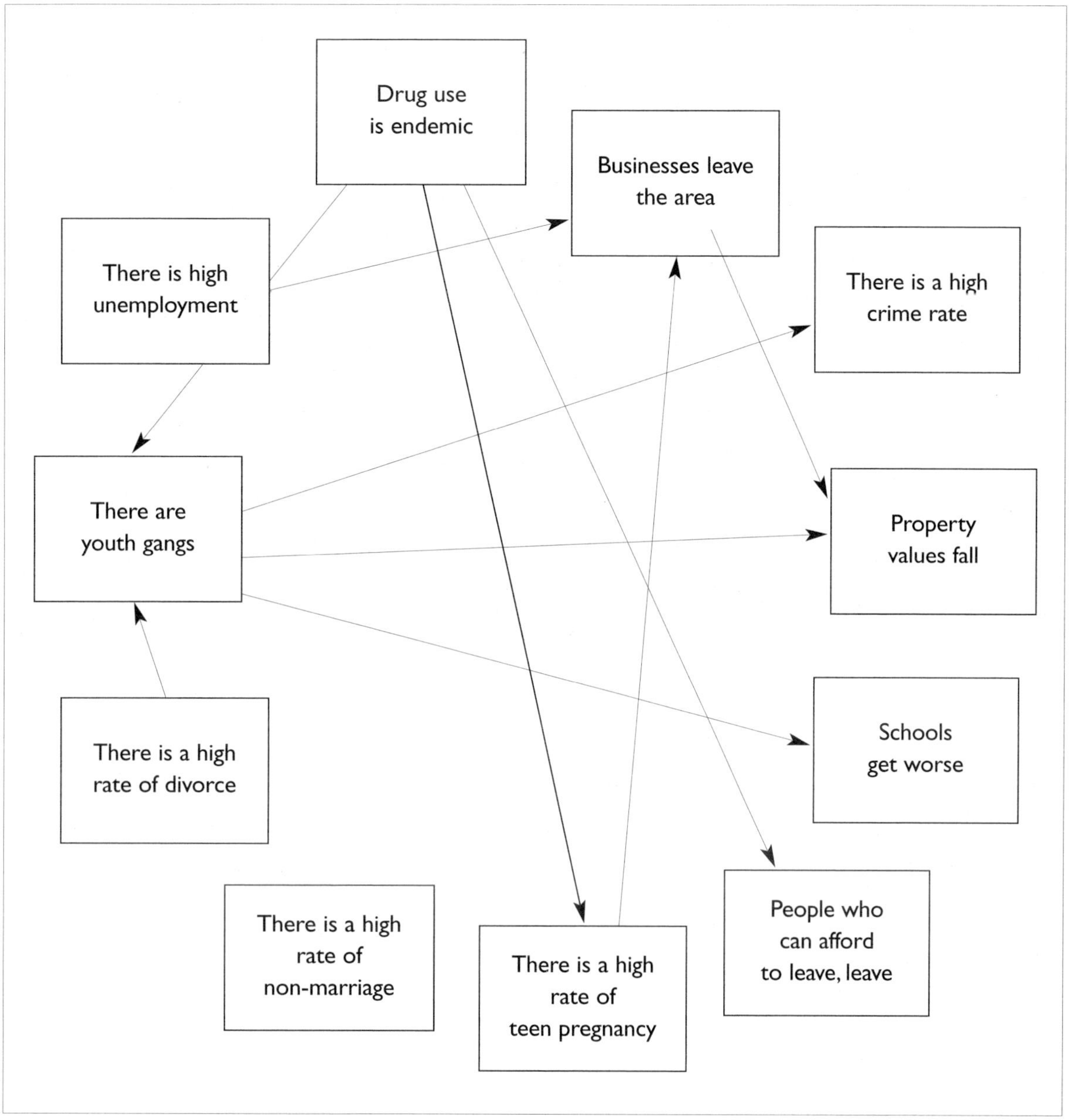

Figure 7-3 Interrelationship diagram.

We can say that one of the reasons for people leaving the city might be the high percentage of drug use there. You can plot a correlation between a high incidence of drug use and youth gangs—the two seem to go hand-in-hand. Because there are gangs, there is a higher than normal crime rate, people start to leave, schools get worse, and property rates nosedive. Many, if not all, of the people who can leave *do* leave, which just exacerbates the situation dramatically.

Obviously, this is not an example that easily fits within the context of the automotive service industry. There aren't many outside the industry that will, and there are fewer still that will fit it from within. However, it isn't hard to overlay our problems on top of the methodology, because that remains constant and will work regardless. We *can* analyze a problem such as having a vehicle or vehicles ready when promised and make our assumptions from there. We can assume that customers don't like it when they have to wait—for a few minutes or a few days! In many cases they have to make arrangements with friends and neighbors for alternative transportation to work, carpools, or even for recreational activities—write such information on a 3 x 5 index card or sticky note. They might have to bear the added expense of a rental car or borrow a car from a friend or relative—another 3 x 5 index card or sticky note.

We can also look forward or backward to the root cause of the problem. We can assume that vehicles are not finished on time for a number of different reasons. We might be doing business with the *wrong* supplier—another 3 x 5 index card or sticky note. We might not be giving the *right* supplier complete, accurate, or adequate information—another 3 x 5 index card or sticky note. Our technicians may not be checking the delivered parts early enough in the process to avert a panic situation at 4:30 PM when it's too late to do anything about it—another 3 x 5 index card or sticky note. While there are situations in which it is impossible to spot additional or ancillary problems until very late in the repair process, it is also possible that our technicians might not be inspecting the vehicles they are working on thoroughly enough to spot potential problems when the vehicle is first brought in to the service bay—goes on another 3 x 5 index card or sticky note. It's possible we could have called the customer to alert them to the problem early enough in the process to avoid any unnecessary inconvenience—if we had the information early enough in the process—perhaps this requires two more 3 x 5 index cards or sticky notes.

As you can see, exploring a problem—dissecting it—can lead to the discovery of countless factors that could or might or should have some impact on the problem. At that point, it's our job to figure out how they all fit together... and, most importantly, what we can do to avoid or at least mitigate the problem when and if we can. The result will be cars and trucks that are finished on time far more often than they would be if we don't.

As you are able to plot the relationships and clarify the ideas, you can begin to focus.

Focus

That focus will result in a clear and compelling statement of your vision, assumptions regarding your personal strengths, weaknesses, opportunities, and threats as well as the strengths, weaknesses, opportunities, and threats facing the company. Classify each issue as either an end result or an action to reach an end result. Group key issues together and then prioritize the strategies that flow out of the exercise, based upon how they move you closer to your vision

SUMMARY

The bottom line is simple. You don't have to remain satisfied with whatever is left over after everyone else takes their *fair share*. You can create a plan and the strategic objectives necessary to make that dream a reality and then pursue it with all your energy and all your might.

What would that plan be? How would you move closer to that dream? What would it take to achieve it?

Take everything we have learned about ourselves, our current reality, our vision, our dreams, and creating a strategic plan and create that plan in the chart on the next page. Write down your vision, your mission, your goals and your objectives, and then write down the steps necessary to achieve that vision, as well.

Thoughts

Actions

Results

CHAPTER 8

References and Resources

As with everything else, there are limits to the amount of time and space available for anything, including this book. Consequently, I have included a list of additional references and resources to help you on your journey.

THE 20-GROUP CONCEPT

The first of these is the concept of the 20 Group that was first established decades ago to help General Motors franchisees establish and share benchmarks and *best practices*. The premise was simple: engage twenty General Motors dealer-principals with dealerships varying in size and revenue—from large to small—from twenty different areas of the country with the express purpose of establishing great performance, and then sharing that information with others in the group.

There are a number of training companies and consultancies providing the same or similar services for independent service dealers across the nation. Although I am sure this list is incomplete, the following is a list of the companies I am personally aware of that provide such a service. If you know of other such companies, please let me know and we will include them as a resource in the next edition.

Automotive Leadership Institute
1204 Brighton Lane
Bel Air, MD 21014
(410) 420-9103
(410) 420-9272 fax
E-mail: info@autoleadership.com

Automotive Service Leaders
Gary Gunn, AAM, CEC
167 Scottsdale Drive
Bowling Green, KY 42103
(713) 639-5385
(800) 233-8551
(270) 782-1707 fax
E-mail: ggunn@prodigy.net

Automotive Training Institute, Inc.
Chris "Chubby" Frederick
8611 Larkin Road, Suite 200
Savage, MD 20763
(410) 792-9466
(888) 471-5800
(301) 498-9086 fax
E-mail: cfrederick@autotraining.net

Bottom Line Impact Groups ®
R.L. O'Connor & Associates, Inc.
Robert "Bob" O'Connor
719 East Main Avenue, Suite B
Puyallup, WA 98372-3306
(253) 841-7878
(800) 755-0988
(253) 840-9834 fax
E-mail: bob@rloconnor.com

Elite Business Services
Bob Cooper
P.O. Box 9630
Rancho Santa Fe, California 92067
(800) 204-3548 inside the US
(858) 756-3102 outside the US
(858) 756-4781 fax
01761 420095 Europe
E-mail: contactus@elitebusinessservices.com

Richard Flint, CSP
Richard Flint Seminars
11835 Canon Blvd., Suite C-105
Newport News, VA 23606-2570
(800) 368-8255
(800) 232-8255 fax
E-mail: Richard@richardflint.com
Website: www.RichardFlint.com

Garage Management System
Doug Kroesche
P.O. Box 90368
Austin, TX. 78709-0368
(512) 301-8356 voice and fax
E-mail: dlkgms@aol.com

NCM Associates, Inc.
10551 Barkley Street
Overland Park, KS 66212
(913) 649-7830
E-mail: info@ncm20.com

COACHING & CONSULTING

I think you will find the same companies mentioned above both willing and able to provide *formal* consulting and coaching services as well. The question then becomes: Do I really need a coach? And, the answer… Well, the answer will more than likely be a resounding Yes!

When the question of including a Reference and Resources chapter came up, a good friend and associate suggested that having a *different set of eyes* on your business was an essential element for success. Having another set of eyes on any element of your performance—in business, in sports, even in your personal life—isn't really all that new and different a concept. It is relatively safe to say that everyone could use an effective coach from time to time, and that *everyone* even includes major league baseball players, NBA basketball players, NFL football players, PGA golfers, professional tennis players, et al. It is equally safe to say that everyone can enjoy the benefit of a little direction, guidance, encouragement, and constructive criticism from time to time as well. So, it really isn't a question of whether or not you could benefit from that other set of eyes—it is really an issue of whether or not you are able to recognize that benefit and then whether or not you choose the right *coach* or consultant.

INSTITUTIONS, AFFILIATIONS, ASSOCIATIONS AND NETWORKS

Just as the idea of coaching, consulting, training, or 20 Groups isn't new, neither are the benefits associated with affiliating with other professionals in your industry. Consequently, I believe with a perfect faith that affiliation with one of the state and/or national associations in our industry will almost always be a positive step—a step in the right direction.

Realistically, there are too many associations to list here. However, it is safe to say that regardless of where you are, finding and contacting the appropriate organization should not prove too difficult. To help you in your quest for information, we are going to include just a few here. They are:

The largest national association of automotive repair, collision and transmission shops is:

The Automotive Service Association
P.O. Box 929, Bedford, Texas 76095-0929
(817) 283-6205
(800) 272-7467
(817) 685-0225 fax
E-mail: asainfo@asashop.org
Website: www.asashop.org

The leading provider of management education for the automotive service industry is:

The Automotive Management Institute
P.O. Box 929, Bedford, Texas 76095-0929
(800) 272-7467, ext. 239
(817) 685-0225 fax
E-mail: info@amionline.org
Website: www.amionline.org

The single largest group of automotive shop owners and technicians interacting online:

I-atn
The International Automotive Technicians Network
41,263 automotive repair technicians (both owners & technicians)
Website: www.i-atn.com

BOOKS

There are countless books on the subject of coaching on both personal and professional levels. There are bookstores filled with books on every aspect of management, operations, sales, and marketing. I know. I've spent hours in such bookstores … and hundreds, perhaps even thousands, of dollars adding to an already unmanageable library. There are, however, a few books that are absolute *must reads*:

The 'E' Myth	Michael Gerber
Flight Of The Buffalo	Ralph Stayer & John Belasco
7 Habits Of Highly Effective People	Stephan Covey
Principle Centered Leadership	Stephan Covey
First Things First	Covey, Merrill & Merrill
Customers For Life	Carl Sewell
How To Win Customers And Keep Them For Life	Michael Leboeuf, Ph.D.
The One Minute Manager	Kenneth Blanchard, Ph.D. and Spencer Johnson, Md
Putting The One Minute Manager To Work	Kenneth Blanchard, Ph.D. and Robert Lorber, Ph.D.
Beware The Naked Man	Harvey Mackay
Swim With The Sharks...	Harvey Mackay
Awaken The Giant Within	Tony Robbins
Guerrilla Marketing	Jay Conrad Levenson
Passion For Excellence	Co-Authored by Tom Peters
In Search Of Excellence	Co-Authored by Tom Peters
Thriving On Chaos	Tom Peters
Liberation Management	Tom Peters
The Pursuit of Wow	Tom Peters
Anything by Tom Peters!	

For those of you with more esoteric tastes:

The Art Of War	Sun Tzu
The Book Of The Five Rings (Spheres)	Miyamoto Musashi
Mastery	George Leonard
The Way of Aikado	George Leonard
The Way of Life	Lao Tzu

As I mentioned previously, my own personal library has expanded to well over 225 such volumes, and I can honestly say that I have learned at least something from each of them—even the *bad* ones! The key here is expanding your mind, thinking outside of the box, and learning how to apply seemingly unrelated principles—principles from other disciplines—to what you and I do.

If I had to pick only one book to recommend to you, it would have to be *The 'E' Myth* by Michael Gerber. It will help you understand the difference between knowing how to do something well and knowing how to run a business that does that same thing well. There is an incredible difference between the two.

Once again, if there is anything—any organization, association, or reference book—that you feel *must* be included here, let me know and we'll see that it is included in the next edition if at all possible.

CHAPTER 9

Conclusion

Cherish your visions and your dreams as they are the children of your soul; the blue prints of your ultimate achievements.

— *Napoleon Hill*

I have an old T-shirt with the following sentiment scribbled across the back:

> *I've been punched, knocked down, kicked around, bitten and beaten… the only reason I'm still hanging around is just to see what happens next!*

That sentiment could easily sum up the cumulative experience of the majority of the people working in the automotive service industry today. We're here and we're working hard everyday. But, we're just hanging out to see what happens next, hoping that whatever it is it won't be—couldn't be—any worse than anything and everything they've already been through.

We are Alice in *Through the Looking Glass,* diligently trying to get to someplace without a clue where that *someplace* ought to be.

It doesn't have to be that way. You don't have to show up for work every day wondering how you will *react* to whatever happens next. There are only two things standing in your way: *fear* and *guilt.* And while there are few of us who have escaped being held prisoner by either or both, it is all too often a self-imposed sentence.

Fear and guilt are two of the most common and most powerful of all human emotional responses. Fear is becoming paralyzed by something that may or may not happen in the future, and guilt is becoming paralyzed by something that has already occurred in the past. Realistically, there isn't much we can do about either except understand what they are and how they impact our lives.

Fear and guilt insinuate themselves into our lives in a particularly insidious way when we are talking about missions, visions, goals, and objectives because they constitute the greatest number of the obstacles we will encounter on our journey. They are insidious because they prevent us from reaching our fullest potential and because, for the most part, they are self-created.

There are countless acronyms for fear, but the two I like the most are False Evidence Appearing Real and Fright Evokes A Reaction. For me, they sum up two of the major problems that accompany fear in general. The first is that because your fears are based on something that may or may not happen, they could easily go unrealized. And, second, fear is reactive. It's hard not to respond to fear—especially to fears associated with something as powerful and compelling as the changes that can accompany success.

Although it may seem counterintuitive, there are far more people terrified of success than there are people frightened of failure. If you fail, you really have nothing to lose! The worst thing that could happen is a return to *the way things were*, which in too many cases is familiar territory. If you fail, all the important things in your life will generally revert back to something comfortable and secure.

Success, on the other hand, is almost guaranteed to bring with it change, and change can be uncomfortable—especially if you are not prepared for it. It forces you to make choices, and with choice comes a loss of opportunity—the realization that by making a decision, you forgo any of the other possible alternatives that might become available to you in the future. Fortunately or unfortunately, that is the price you pay for making a decision and taking action!

To achieve your goals and objectives, to create a more positive, more attractive alternative for the future, or to achieve something extraordinary you must overcome your fear of success.

It isn't really all that complicated, nor is it particularly difficult. It just takes a little bit of planning—a couple of minutes every day just thinking about the future and your place in that future. It isn't any harder than planning a vacation, a trip to the beach, or a weekend away. It takes a point of origin and a destination, that's all. The rest is just a matter of logistics—what will we need to make the journey, who will be coming with us, and how will we get there. We just tend to make it seem much harder—much more difficult than it actually is.

You don't have to accept what's left over—after everyone else gets *their's*—as your portion. It doesn't have to be that way. You don't have to accept the fact that this is as good as it gets because you've never experienced *better*. You can start out each new day, each new week, each new month, each new year with a plan—realizing that your success does not need to be judged by whether or not you achieve all of the goals and objectives within that plan. You can be judged successfully on the basis of how close you come.

It doesn't have to be complicated. It doesn't even have to be all-encompassing in the beginning. In fact, it can be a fairly modest plan focusing on only one or two areas of your business. But you do have to spend at least a little time thinking about alternative outcomes—new and different realities.

So, take what you've learned here and put it to work for you. Craft a vital and vivid image of a different reality—a better reality—and then consider what it would take to make that reality real! Think about the principles that guide you through the minefields of life and business. Write down your vision of the future. Develop the mission, goals, and objectives necessary to make your mission come alive, and then fully commit yourself to the tactics, goals, and objectives necessary to drive yourself into a new and better future.

It's all a matter of risk, reward, and the answer to one simple question: *What have you got to lose?*

If you always do what you always did, you'll always get what you always got!

You've already got a pretty good handle on what you've got. If it was *good enough*, you probably wouldn't have gotten this far in the text! The fact that you have read this far indicates that you think what you've got could be improved, even if only a little.

In too many cases, the success you see others enjoying in the present didn't happen by accident. It was planned. The future you've always dreamt about—the one you see others enjoying right now—could be yours.

Give it a shot! You deserve it. Besides, what have you got to lose? The worst thing that could happen is a tomorrow that looks a little too much like yesterday! Isn't it time for something just a little bit better—a little different?

Just remember...

It is not the strongest species that will survive, nor the most intelligent, but the one most responsive to change.

Charles Darwin

MORE TO COME

This series—*Automotive Service Management*—is divided into three general areas of concentration and eight volumes. As we make our way through these eight volumes, we will be looking at virtually everything involved in successfully operating a contemporary automotive service facility in today's dynamic and challenging marketplace. That means we will be looking at everything from hiring practices, productivity, and service management to the *cost of doing business* (factored for service bay productivity), cost/value ratios, value-added service, and hazardous materials communication. As well as exploring our relationship with the consumer, our suppliers, our employees, and our industry, we will also be looking at promotions, advertising, sales, and marketing.

Our goal is to create the most comprehensive resource for automotive shop management available anywhere—a complete and exhaustive look at all the *hard* and *soft* skills required to manage any business, as well the tools required to run an efficient, effective, customer-responsive automotive service facility.

The series was not conceived or formulated in a vacuum. It did not originate in theory first and then find its way into reality. It is the result of a lifetime spent in the automotive industry backed up by four generations of genetic memory. I invite you to take whatever you can from these pages. Take the KPIs you find here and create a *snapshot* of your business, as it is at the beginning of this journey. Then, see how integrating what you find within the pages of these manuscripts changes those numbers... how it impacts the bottom line—your bottom line. Then, let us know how you're doing!

In addition to this text, there are seven other titles for you to read, enjoy, and work through, including *Total Customer Relationship Management, Managing Dollars with Sense, Operational Excellence, Building A Team, The High Performance Shop, Safety Communications*, and *Operations Management.*

If you have any suggestions for subsequent subjects or ways we can improve what we have already presented, let me know. My goal is to provide you with answers to the questions I had to search for or create myself. My goal is to insure that you don't find yourself struggling the way we had to. My goal is to see that you are richly rewarded for your efforts and your ability. My goal is to see you succeed where others have failed.

So, read this shop operations series one section at a time, one guide at a time, or all at once. See what you like. See what you aren't all that comfortable with. *Then, read it again!* Find those principles that make sense to you. Create a plan to integrate them into your business. *Then, read it again!* Make the changes required to take your business to the next level. Monitor your progress. Document everything... *Then, read it again!*

If it works for you, let us know. If you can find ways to help us help you make it work more effectively, let us know that, too!

REFERENCES

The American Heritage Dictionary of the English Language, 3rd ed.

Barker, Arthur. *Future Edge*. New York: William Morrow & Company, 1992.

Clancy, Kevin J. and Krieg, Peter C. *Counterintuitive Marketing: How Great Results Come from Uncommon Sense*. New York: Free Press, 2000.

Covey, Stephen. *The 7 Habits of Highly Effective People*. New York: Simon & Schuster, 1990.

Drucker, Peter. *Managing In a Great Time of Change*. New York: Dutton/Plume, 1997.

Jennings, Jason, and Haughton, Laurence. *It's Not the Big That Eat the Small, It's the Fast That Eat the Slow!* New York: HarperCollins, 2001.

Peters, Thomas and Waterman, Robert H. *In Search of Excellence: Lessons From America's Best-Run Companies*. New York: Warner Books, 1988.

Senge, Peter. *The Fifth Discipline: The Art and Practice of The Learning Organization*. New York: Doubleday & Company, 1994.

Smith, Adam. *Powers of the Mind*. New York: Ballantine Books, 1976.

#26